THE PRIVATE JOURNAL
OF CAPTAIN G.H. RICHARDS

The Private Journal of Captain G.H. Richards

THE VANCOUVER ISLAND SURVEY (1860–1862)

EDITED BY

Linda Dorricott &
Deidre Cullon

RONSDALE

RONSDALE PRESS
3350 West 21st Avenue, Vancouver, B.C.
Canada V6S 1G7
www.ronsdalepress.com

Typesetting: Julie Cochrane, in Granjon 11.5 pt on 14.5
Cover Design: Julie Cochrane
Front Cover Images: Captain George Henry Richards (BC Archives A-02432); H.M.S. *Plumper* (BC Archives Pdp 00076)
Frontispiece: Captain George Henry Richards (BC Archives A-03352)
Paper: Ancient Forest Friendly "Silva" (FSC) — 100% post-consumer waste, totally chlorine-free

Ronsdale Press wishes to thank the following for their support of its publishing program: the Canada Council for the Arts, the Government of Canada through the Canada Book Fund, the British Columbia Arts Council, and the Province of British Columbia through the British Columbia Book Publishing Tax Credit program.

Library and Archives Canada Cataloguing in Publication

Richards, George Henry, Sir
The private journal of Captain G.H. Richards: the Vancouver Island survey (1860-1862) / edited by Linda Dorricott & Deidre Cullon.

Includes bibliographical references and index.
Issued also in an electronic format.
ISBN 978-1-55380-127-6

1. Richards, George Henry, Sir — Travel. 2. Richards, George Henry, Sir — Diary. 3. Vancouver Island (B.C.) — Description and travel. 4. Indians of North America — British Columbia — Vancouver Island. I. Dorricott, Linda II. Cullon, Deidre III. Title.

FC3844.R53 2011 971.1'2042 C2011-903278-3

At Ronsdale Press we are committed to protecting the environment. To this end we are working with Canopy (formerly Markets Initiative) and printers to phase out our use of paper produced from ancient forests. This book is one step towards that goal.

Printed in Canada by Marquis Printing, Quebec

In memory of
Jean-Louis Brachet
and Ed Sanders

ACKNOWLEDGEMENTS

This journal has been made available to the general public with the permission of the owner, Donal Channer. Donal and Jill Channer have been with us all the way providing insight, encouragement and a strong sense of historical connection to Captain George Henry Richards. We have enjoyed many hours of animated conversation around the dinner table and in the library of their charming home in Wiltshire.

Among the First Nations that have supported our work over the years, we would like to particularly thank the Laich-Kwil-Tach, the Quatsino and the Gwa'Sala 'Nakwaxda'xw.

Much of our research in England took place at the Hydrographic Office in Taunton where archivist Guy Hannaford has educated and assisted us since 2006. We also spent long hours at The National Archives at Kew, the British Library and the Cambridge University Library where we were continually impressed by the efficiency of the service.

We wish to acknowledge the staff at the Royal British Columbia Museum and the British Columbia Archives in Victoria, particularly Dan Savard and Kelly-Ann Turkington for their assistance with the photographs. We also thank the staff at the Vancouver Maritime Museum, the Vancouver Island Regional Library, Vancouver Island University Library, and Special Collections and Archives at the University of British Columbia.

Friends, colleagues and family all worked to make this publication possible. Alan Twigg was instrumental in taking us from the idea of publication to its realization. David Scott donated his time to develop the Vancouver Island maps. Jean-Louis Brachet, Sandro Brachet and Arlene Sanders assisted in research and dedicated many hours to deciphering and transcribing handwritten documents.

Finally we would like to thank our publisher Ronald Hatch for his patience and his enthusiastic support for our project, the designer, Julie Cochrane, and the editors at Ronsdale Press who have helped to make our book a reality.

CONTENTS

Preface

In the spring of 2006 the editors travelled to England in search of the original surveys, logs, field notebooks, journals and letter books of Captain George Henry Richards. We were eager to review these documents on behalf of our First Nation clients, hoping to find material that would add to existing ethno-historical information and provide support for land and resource claims. We were particularly interested in Richards' journal of the Vancouver Island surveys conducted between 1860 and 1862.

Our first stop was the Hydrographic Office (the UKHO) in Taunton, where archivist Guy Hannaford gave us ready access to all of the original charts and drawings of the Vancouver Island surveys. Twenty-first century satellite photography cannot render the misty seas and immense forested shores suggested in these delicately etched and colour-washed nautical drawings. The published version of these charts can be viewed at the British Columbia Archives in Victoria, but certain details of the original drawings have been omitted and our interest was in these omissions, particularly the barely visible rectangular boxes on the shore which represent the houses and villages of many First Nations on Vancouver Island.

At the UKHO we were directed to a cottage near Westbury, Wiltshire, where Captain Richards' manuscripts have pride of place in the private library of Donal Channer, the great-great-grandson of George Henry Richards. The collection came into the possession of the Channer family through Captain Richards' Victoria-born daughter, Rose, who married Arthur Channer. Donal Channer now has responsibility for its preservation.

At the Channer home, along with tea and hospitality, we were given permission to photograph all of Richards' Vancouver Island collection. In a colourful bound journal in excellent condition entitled *Vancouver I^{d} Survey H.M.S. "Plumper" 1860. Captain's Journal GHR*, the story of the numerous circumnavigations and surveys of Vancouver Island is recorded in full and vivid detail. Upon reading the journal and recognizing the contribution it would make to our knowledge of the early colonial period on Vancouver Island and in British Columbia, we requested and were given permission by Mr. Channer to publish the journal.

The journal is presented here in its entirety with as few editorial changes as possible. The text is liberally annotated and we have supplemented Richards' account with excerpts from the journals of John Thomas Ewing Gowlland, Richards' second master. Described by Richards as "a most competent surveyor and along with Master Browning, his best draughtsman," Gowlland wrote at least as many pages as Richards, providing additional detail and bringing a youthful, though less-tempered, perspective to his account. His attitude toward the aboriginal inhabitants is certainly not as respectful as Richards' and he exhibits a Victorian prejudice that may not be considered acceptable to the modern reader.

Richards' trusted officer and friend, Lieutenant Richard Charles Mayne, also kept a record of the Vancouver Island survey expedition until 1861 when he was promoted to his own command and left the West Coast to serve in New Zealand. Mayne's journal was incorporated into a book well-known to local historians, entitled *Four Years in British Columbia and Vancouver Island*. The original journal and typescript of Mayne's journal are available at the British Columbia Archives and his book is available online. While Mayne's *Four Years* has been mined extensively, historians of Vancouver Island have paid less attention to Gowlland's lengthy and detailed journal. Gowlland's journal is held by the Mitchell Library in Sydney, Australia, and copies of the handwritten journals are available on microfilm at the British Columbia Archives and at the University of British Columbia Library. Together, the journals of Richards, Mayne and Gowlland give us a balanced and complete version of one of the most significant survey expeditions on Vancouver Island.

The publication of the *The Private Journal of Captain G.H. Richards: The Vancouver Island Survey (1860–1862)* is long overdue. It is intended to recognize Richards' official accomplishments as well as the personal qualities of balance, tolerance, integrity and perseverance that are his legacy to British Columbia.

— Linda Dorricott & Deidre Cullon
Nanaimo, 2012

Introduction

Anyone who has sailed the coastal waters of Vancouver Island is familiar with Pender Island, Mayne Island, Gowlland Harbour, Bull Passage, Browning Inlet, Blunden Harbour, Bedwell Harbour and Mount Moriarty. What they are less likely to know is that these geographical and nautical features have one thing in common: they are all named for the officers and seamen of the Royal Navy who served under Captain George Henry Richards between 1857 and 1862 on two naval survey ships, the H.M.S. *Plumper* and the H.M.S. *Hecate*.[1] In these few years they sounded, sailed and charted the entire coastline of Vancouver Island and much of the mainland coast of British Columbia, creating the baseline information for the nautical charts that we use today. Every ship that sails these treacherous waters and finds safe harbour owes a debt of gratitude to Captain Richards and his crew.

Much has been written about the men who mapped the lands and waters

[1] Plumper Pass, Plumper Sound and Hecate Strait take their names from these two survey ships.

of what is now the province of British Columbia. Numerous books, articles and journals describe the lives and exploits of such men as Captain George Vancouver, Colonel Richard Clement Moody of the Royal Engineers, Colonial Surveyor Joseph Despard Pemberton, Robert Brown of the Vancouver Island Exploring Expedition and George Dawson of the Geological Survey of Canada. But little is known of Captain George Henry Richards and the role he played in the early development of the colonies of Vancouver Island and British Columbia.

Never the focus of any major published work, Richards and his significant contribution as a surveyor and chart maker have not been widely recognized. Andrew S. Cook provides the most comprehensive overview of Richards as a surveyor and chartmaker.[2] Historian Barry Gough situates him in the context of the naval history of the West Coast, describing his role in the British American boundary dispute and in law enforcement on the Fraser River and along the inside passage between Vancouver Island and the mainland.[3] Richards' ships and officers also make appearances in G.P.V. and Helen B. Akrigg's *British Columbia Chronicle, 1847–1871*.[4]

Captain John T. Walbran frequently refers to Captain Richards in his comprehensive *British Columbia Coast Names, 1592–1906*. Although Captain George Vancouver was responsible for naming the major land and water features of Vancouver Island, including many of the names given by

---

[2] See Cook's *The Publication of British Admiralty Charts for British Columbia in the Nineteenth Century*. A comprehensive account of Richards' survey expedition can also be found in *Charting the Northwest Coast: 1857–1862*, an unpublished master's thesis by Richard William Wallace.

[3] See Gough's *The Royal Navy and the Northwest Coast of North America, 1810–1914* and *Gunboat Frontier*.

[4] The Akriggs obtained a copy of Richards' journal and his letterbook to the Hydrographer in the early 1970s. Helen Akrigg also references the Channer Collection in her entry in the *Dictionary of Canadian Biography*, stating that the collection "remains in private hands in England. The author was permitted to consult these papers but is not at liberty to disclose their location." There has been a certain amount of secrecy surrounding these documents and a reluctance on the part of researchers to share them. In fact, photocopies of the journal and letterbook are in the Akrigg Papers at the Archives of the University of British Columbia Library. A typescript of the journal prepared under the Akriggs' supervision is available at the Vancouver Maritime Museum. According to Donal Channer and Guy Hannaford, archivist at the UKHO, no copies have been deposited in British archives.

such Spanish explorers as Dionisio Alcala Galiano, Cayetano Valdes and Juan Francisco de la Bodega y Quadra, Richards, charting in much greater detail, provided the nomenclature for the majority of the smaller features: islets, small bays, inlets and harbours, points and rocks. He also adopted native names whenever possible, particularly on the west coast of the Island: among them are Quatsino, Klaskish, Kyuquot, and Ahousat (Sandilands 1983: 3).

From the time of his arrival on Vancouver Island in November 1857 to his departure for England in December 1862, Richards reported to three masters: his commanding officers, the Crown Colony of Vancouver Island and the Hydrographic Office of the Admiralty. His duty to each resulted in a large body of reports and correspondence.

His reporting letters to his superior officers, the commanders-in-chief of the Pacific Station, Rear Admiral Sir Robert Lambert Baynes (1857–60) and Rear Admiral Sir Thomas Maitland (1860–62) are mainly concerned with naval matters and questions of military security on the Fraser River and in Georgia, Johnstone and Queen Charlotte straits.

His correspondence with Governor James Douglas and his administrative assistant, Colonial Secretary William Young, illustrates the wider range of duties imposed on him by the colonial government and is a rich source of information on the early infrastructure of the two colonies of Vancouver Island and British Columbia. Richards wrote lengthy reports and proposals on harbours, lighthouses and buoys and made recommendations for agricultural settlement lands and government reserves, including military and Indian Reserves. The colonial correspondence also provides early information on timber, minerals and other natural resources as well as details of overland expeditions made by his crew on the mainland and across Vancouver Island.

Richards' correspondence with Rear Admiral John Washington, hydrographer of the navy, provides the most complete record of Richards' sojourn on Vancouver Island. Then, as now, the Hydrographic Office was responsible for the production and publication of charts for the Royal Navy. The letterbook held in the Channer Collection is a fair copy of the correspondence from Richards to Washington from 1857–1862. Washington's instructions provide the framework for all of Richards' activities, and his reporting letters detail the progress of the survey and chart-making work.

Until now, these official sources have formed the basis of the information in the published record. The publication of the journal reveals the man behind the official reports and breathes life into a little-known period in the history of Vancouver Island.

AN OFFICER AND A GENTLEMAN

Born in 1819[5] to Captain George Spencer Richards in Antony, on the coast of Cornwall, George Henry Richards entered the navy at the age of thirteen. At fifteen he was appointed to his first surveying duty in the Pacific on the *Sulphur* under the command of Sir Henry Belcher. Richards' first visit to Nootka Sound on the *Sulphur* in 1837, and his encounter with Chief Maquinna, made a lasting impression on the young midshipman, exciting an interest and a respect for the aboriginal inhabitants of Vancouver Island that is displayed throughout the pages of his journal. It is fortunate that among his other accomplishments, Richards was a careful observer and, from an early age, a journal keeper. According to Donal Channer, Richards' earliest journal starts in August 1838 and runs to April 1840. In it "Richards describes a visit to San Francisco, his first command surveying in the pinnace *Victoria* 'under my own pennant' and he ends by remarking that he is going to find more paper to make another book in which to 'write more nonsense for the amusement and edification of God Knows Who'" (Channer 2008, personal communication).

Richards' early career combined active military service as well as surveying duty and he distinguished himself in both services. He was a natural leader: brave and clearheaded in battle and conscientious and driven to excellence as a surveyor. The official biographies are full of references to his bravery, his zeal and his exemplary conduct. He saw active duty during the first Opium War with China in 1839, and as second lieutenant and assistant surveyor on the *Philomel*, he surveyed the Falkland Islands and the southeast coast of South America. In 1845 he was appointed to commander after leading his men in military action against the Republic of Buenos Aires.

Sir Edward Belcher, a captain known to be "exacting and difficult to

---

[5] Although most official biographies give 1820 as Richards' year of birth, the editors have presumed that the baptismal date of February 27, 1819, in Akrigg's biographical note is more accurate (Akrigg, 2000).

please," described Richards as having "at all times borne the character of an exemplary and steady officer, and is one of the few officers of the *Sulphur* of whom I can speak with unqualified praise, not only for his assiduity in surveying, but for his gallantry during the operations at Canton, and for his exemplary conduct when the other officers of the *Sulphur* were in a state of insubordinate alienation from their Captain" (Dawson 1885: 135).[6]

Richards' bravery was not limited to military action. For four years as second captain and assistant surveyor on the New Zealand surveys he "was often much exposed when detached, carrying the more detailed or laborious portion of the work, in open boats." Following the completion of the arduous New Zealand survey and the compilation of the *New Zealand Pilot*, Richards volunteered to serve as second-in-command under Belcher, sailing to the Arctic in a fruitless search for the missing expedition of Sir John Franklin. Belcher's behaviour on this expedition has been described as "more overbearing and unreasonable than ever," and Richards' "tact and judgement were critical in holding the operation together." He nevertheless found time to complete a ninety-three-day sledge journey considered to be one of the most extraordinary on record. Upon his return, Belcher was court-martialled for abandoning four of his ships in the Arctic. Richards kept a private journal of this Arctic expedition with the melodramatic introduction: "If any person ever makes public the writings in this diary may he be haunted by my ghost in this world and the next" (Ritchie 1995: 261).[7] Fortunately Belcher was acquitted and Richards' diary remains private.

Between 1857 and 1862 Richards served the colonies of Vancouver Island and British Columbia as boundary commissioner, military "peacekeeper" and surveyor. His short but influential time on the West Coast was to be his last term of active surveying duty, and upon his return to England in 1863 he took up the prestigious post of hydrographer of the navy, filling the position vacated by the death of Rear Admiral Washington. Over the next ten years, Richards sought to modernize the Hydrographic Department through numerous innovations aimed at consolidating and publishing

[6] Unless otherwise cited, all biographical quotations are from Dawson 1885: 134–136.

[7] Ritchie suggests that Richards kept a record of events in case Belcher later court-martialled his officers "as was his habit." According to Donal Channer, Belcher was exonerated of wrong-doing but did not receive an absolute acquittal even though "his sword was returned" (Channer 2008, personal communication).

hydrographic information and making it available for general use. He also promoted oceanographic research and was a motivating force behind the three-year scientific expedition of the H.M.S. *Challenger*, an expedition which he described as "the hope and dream of his life."

After his resignation as hydrographer in 1874, Richards became managing director of the Telegraph Construction and Maintenance Company where, under his direction, 76,000 miles of submarine cables were laid. Between 1870 and 1884, Richards was promoted from rear admiral to vice admiral to admiral. He was knighted, received the Order of Bath, and became a fellow of the Royal Society and the Royal Geographical Society and a member of the Academy of Science of Paris.

Little is known of Richards' private life. In March 1847, he married Mary Young, the daughter of Captain R. Young of the Royal Engineers. Mary arrived on Vancouver Island sometime in 1858 with an infant girl, and in the two years she spent in Victoria, she gave birth to another daughter and a son. In November of 1860, Richards records that on "the evening previous a son was born to me — and his name was called Vancouver" (Richards, November 1860). This is one of the rare references he makes to his family in the journal.

We learn from Lieutenant Mayne that on Christmas Day in 1860, they "dined at Capt. R's, smoked and sang songs till a late hour as is usual on such occasions — this is our fourth Xmas together and last in the poor old 'Plumper' — the last too for Mrs Richards out here — the first one I remember we all dined in the Chart room, the 2nd on the Quarter deck — Mr & Mrs Crickner with us, the 3rd as this one, at the house — where shall we all spend our next?" (Mayne 1857–1860: December 25, 1860). The house referred to was probably Thetis Cottage on Esquimalt Harbour where Richards lived with his family, close to where the *Plumper* was anchored. Mayne adds that Mrs. Richards will be returning to England on the *Princess Royal* and he "does not suppose the Captain will be anxious to stay after he has served his time" (Mayne, December 31, 1860).

Gowlland, too, provides some insight into the relationship between Richards and his wife. He writes that the crew have gone to Esquimalt "but I imagine they will not have long to Enjoy the sweets of that city; as the great tie that attracted us to that place has disappeared in the Princess Royal to England; and the Captain is most anxious to get as great a quantity of Coast line done this year as possible toward expediting our return to England in 1863 please God to spare us long enough" (Gowlland 1860–1863: June 1861).

A photograph of Captain Richards and his officers shows Mrs. Richards front and centre, an unusual inclusion in a portrait of naval officers. Donal Channer observes that the presence of Mrs. Richards is "surprising" and "would have required special permission which was not always granted" (Channer 2008, personal communication).

Back in London in 1863, Richards found his new career as hydrographer demanding, and although he excelled as an administrator, as he did in all his endeavours, it may be that he had left behind a life better suited to his temperament. He writes from the Hydrographic Office, "This place like all permanent places in these times is full of work and I get no leisure. I often wish myself at sea again. My wife is not strong and is very often a great invalid. Children all well however" (Richards 1863).

Eleven years later, in his letter of resignation as hydrographer, he writes:

> A long and unbroken service afloat of a very rough character admonishes me that the unceasing duties of a department such as this is 'telling' upon my health which in the interests of a large family of children I must endeavor to guard. There are numerous other reasons of a personal nature. (Richards 1873: December 19)

He gives "family reasons" for accepting the new position of managing director of the Telegraph Construction Company, describing his position with the Admiralty as "a very Wearing occupation and more like a Commercial place than a Gov. dept. [that] has always been a losing concern to me. I may as well accept the actual commercial position with the real remunerative conditions attached to it" (Richards 1874: May 9). Presumably Richards' "family" reasons included financial reasons. The 1881 census shows Richards supporting a household of twelve, including his wife Mary who died that same year, two sons, four daughters, one grandson and four servants plus one visitor. The year after Mary Richards' death, Richards, sixty-two, married Alice Mary Tabor.

Richards died of complications from sciatica on November 14, 1896, while taking the waters in Bath (*The Times* 1896: November 17). His obituary described him as "a man of great ability, of sound commonsense, and of untiring activity, and his unfailing good humour, general shrewdness, and kindness to young members of his profession caused him to be universally loved and respected."[8]

RICHARDS ON VANCOUVER ISLAND

When Captain Richards arrived in the Colony of Vancouver Island in November 1857 he brought a global perspective to the job. After twenty-five years at sea and two circumnavigations of the world, he had proven himself in battle and in the fearless exploration of dangerous and uncharted waters. He was an accomplished surveyor and chart maker with a growing interest in the new science of oceanography and a dogged determination to make the oceans safe for seafarers. These achievements, along with his ability to command the respect of both his superiors and his men and to exercise diplomacy and judgement in the face of conflict, made him the ideal person to execute the diplomatic, military and hydrographic requirements of a colony undergoing rapid transformation.

THE BOUNDARY COMMISSION

> On arrival at Vancouver's Island your first duty will be to consult with Captain Prevost as First Commissioner on the provisionary steps to be taken to prepare and to define a portion of the Boundary lines between H.M. Possessions in North America and the United States 1st Article of the Treaty of Washington of the 15th June 1846 as fully set forth in your General Order from their Lordships. And when you shall have received his Instructions you will lose no time in setting out an accurate nautical survey of such portions of the channels and islands which lie between Vancouvers island and the Continent of America so as to obtain if possible in the autumn of the present year such information as may be required for a due consideration of the whole question as soon as the Commissioners on the part of the U.S. shall be ready to enter into the discussion. (Washington 1857: March 16)

Captain George Henry Richards arrived at Esquimalt Harbour on November 10, 1857, on the H.M.S. *Plumper*, following his appointment as second British commissioner and chief surveyor and astronomer to the British

[8] Obituary in the *Proceedings of the Royal Society*, cited in Akrigg, 2000.

Boundary Commission. A clear and well-defined line had been drawn along the forty-ninth parallel, neatly dividing the British and American territories, but before the final lines were drawn it became apparent that where the forty-ninth parallel meets the coast, the existing marine surveys did not adequately describe the waters of the several straits between southern Vancouver Island and the American mainland. Richards' instructions were to put this nautical disorder to rights.

The Boundary Commission was first established in 1856, and in 1857 Captain James C. Prevost was appointed the first British boundary commissioner. While Prevost and the American commissioner, Archibald Campbell, wrangled over the location of the maritime boundary, Captain George Henry Richards and his crew systematically imposed order on the chaos of the coastal waters of what is now the province of British Columbia and the northern coast of the state of Washington. According to the Akriggs, the dispute may well have been better settled through the diplomacy of Richards and his American counterpart, Lieutenant Parkes, than by Prevost. They describe Captain Richards as "rather a small man physically but a mass of energy, shrewdness and humour, as strong as Prevost was weak" and suggest that if the boundary issue had been left to the skill and diplomacy of Richards and Parkes, a compromise would probably have been reached to situate the boundary down a "Middle Passage" and today San Juan Island would be part of British Columbia (Akrigg 1977: 101).[9]

SURVEYS AND CHARTS

> As soon as the work connected with the Boundary Commission shall be completed & Capt. Prevost shall have no longer occasion for your services you will proceed with the survey of the Gulf of Georgia & the harbours of Vancouver Island according to their importance. Of this you will be a better judge on the spot than any one here. You will of course be guided by the discovery of coal & other facilities for the supply of our ships. You will not fail to send [illegible] tracings of all surveys and places and copies of all descriptions &

[9] The stalemate over the San Juan Boundary led to an escalation of Anglo-American tension, culminating in a military standoff between Governor James Douglas and American Brigadier General W.S. Harney. See Gough 1977: 156–163 for an account of the Royal Navy's role in what came to be known as the "Pig War."

> Sailing Directions in order that when expedient they may immediately be communicated to Lloyds and made public for the benefit of Sailors in general. You should also communicate any important facts to the Cf [Commander in Chief] on his Station and to the Authorities on the spot so that it may at once be made available for navigation. (Washington 1857: March 16)

During his five-year commission on the West Coast and guided by the charts of his predecessors including George Vancouver, Aemilius Simpson, Edward Belcher, Henry Kellett, James Wood and G.H. Inskip (Cook 2004: 51–55), Richard completed surveys of the Strait of Georgia and the Gulf Islands, the Fraser River up to Fort Langley, the entire coastline of Vancouver Island, Burrard Inlet, Howe Sound, Jervis Inlet, Bute Inlet and the harbours of Victoria, Esquimalt and Nanaimo. He completed thirty-six principal charts of the coast and in 1861 and 1864 compiled and published the first two editions of the *Vancouver Island Pilot*, the first complete set of sailing directions for Vancouver Island and the south coast of the mainland.[10] Between 1863 and 1865, in his capacity as hydrographer, Richards directed the remaining surveys of coastal British Columbia. These surveys were carried out by Richards' master, Captain Daniel Pender on the Hudson's Bay Company vessel, the *Beaver*. Pender's survey work was incorporated into the *Vancouver Island Pilot* and updated as the *British Columbia Pilot* in 1888. Using this guide alone, a sailor today could still safely navigate the coastline of British Columbia.

Captain Richards' journal is an account of three of these survey years: 1860, 1861 and 1862. Most of the 1860 surveys were completed on the east coast of Vancouver Island from Cape Lazo to Cape Mudge, Johnstone and Broughton Straits to Queen Charlotte Strait, Jervis Inlet, Howe Sound and Texada and Lasqueti Islands. A partial examination was also made of Quatsino Sound. During the 1861 and 1862 seasons Richards and his crew were primarily engaged in the surveys of the west coast of the Island from Goletas Channel to Barkley Sound.

Survey work was conducted between early spring and late fall with a considerable amount of to and fro up and down the Island; the ships did not circumnavigate in an orderly fashion. During the 1860 season alone

[10] See Dawson 1885: 139 and Cook 2004: Table 3:1 for a complete list of admiralty charts.

the H.M.S. *Plumper* left for Nanaimo in April, surveyed the east coast as far as Beaver Harbour, returned in June to Esquimalt, was called upon to settle a dispute in Fort Rupert in August and continued north to Hope Island and Bull Harbour. She then proceeded south along the west coast of the Island to Quatsino and Nootka Sounds returning to Esquimalt in September. In October she returned to Nanaimo, travelled up Georgia Strait to Jervis Inlet, spent November on the Fraser River, returned again to Nanaimo and finally returned to base in Esquimalt in December. Nanaimo was a frequent port of call, and poor service at the coaling station was the cause of much grumbling and complaint.

The ships wintered at Esquimalt where the officers and crew were primarily engaged in chart work and ship repairs with occasional leave granted to officers and crew to take part in the social events on offer in Victoria. Officers and crew were also engaged in such duties as erecting boundary markers for the Boundary Commission, setting buoys in channels and conducting overland exploratory missions.

THE SURVEY SHIPS

The Vancouver Island Surveys were conducted on two survey ships: the H.M.S. *Plumper* and the H.M.S. *Hecate*. The *Plumper* was a 484-ton, 60-horsepower steam sloop barque rigged and "armed with two long 32 pounders and ten short ones" (Mayne 1862: 10). Although a chart room had been built on board, the lack of space and light made it necessary to do chart work on shore at Esquimalt during the winter season. By the end of 1859, after three survey seasons in Haro and Rosario straits, the Fraser River, Burrard Inlet, Victoria, Esquimalt, Nanaimo and Comox, it had become clear that the *Plumper* would not be adequate to meet the challenges of the Vancouver Island surveys. In a letter to the hydrographer, Richards outlined some of the deficiencies of the *Plumper*:

> Various difficulties begin to take place when a vessel has been 4 years in commission. The first of these is the men, a great number of whom will have completed their engagements; it is impossible to replace them here as proof of which I have been from 10 to 14 short, ever since our arrival and have never been able to fill up although the whole squadron has been present. The second is, the Machinery itself boilers etc, will begin to complain although up to this time I must say, has been kept in excellent order with

> constant wear. Lastly, I must candidly say although the Plumper has a vast advantage, over a sailing vessel, yet she is by no means well adapted for surveying in this country her maximum speed is 6½ knots. In another year our work will be in the narrow straits of the N.E. end of the Island where the tides run commonly 8 knots. . . . (Richards 1857–1862: December 21, 1859)[11]

In the same letter Richards requested a replacement ship, preferably a paddle sloop with good accommodation, height, light and a chart room big enough for a staff of six to eight. This request was granted, and in December 1860, H.M.S. *Hecate* arrived in Esquimalt Harbour. It was an 810-ton, 240-horsepower Symondite paddler with new boilers and a crew of 125 officers and men. Gowlland noted she was "rather an old craft built in 1839 — but by no means ugly. . . . She is a most roomy comfortable ship and practically adapted to surveying purposes. Our fire Room is quite a palace compared to the one we have just left; and we can dine 12 at the table without any inconvenience" (Gowlland 1860–1863: January 1, 1861).

Even the *Hecate*, however, was no match for the treacherous coast off southern Vancouver Island. During her first survey season, in August of 1861, the *Hecate* hit rocks in a night fog in Juan de Fuca Strait. Extensive damage to her hull led to a two-month hiatus in dry dock at Mare Island in San Francisco Harbour, the only dock on the Pacific coast where repairs could be made. During this period, an impatient and often disapproving Richards is clearly out of his comfort zone, but nevertheless manages to fill the pages of his journal with a humorous and lively account of his adventures in the lawless "Wild West" during the politically charged atmosphere that immediately preceded the American Civil War.[12]

---

[11] Donal Channer provides an interesting detail about the *Plumper*. "The *Plumper* was driven by sail and steam-powered screw. The problem with the screw is that it slows the ship down and makes it difficult to steer when under sail. . . . To overcome this problem, the *Plumper* was so arranged that the screw could be removed and hauled up through the ship — including Richards's cabin — and stored on deck (Channer 2008, personal communication).

[12] While Richards was in San Francisco he took the opportunity to examine the state of the defences of the San Francisco Harbour and determined that in the case of hostilities, "the city itself, would have been at our mercy, as also the works at Mare Island." He is less confident about the ability of the Colony to defend itself (Richards 1857–1862: February 17, 1862).

The journal also describes the importance of the survey work performed in small boats dispatched from the main ship. The journal mentions no less than nine support boats: the *Shark*, the pinnace, the galley, three whalers, the gig, the cutter and the dinghy.[13] The survey crews that manned these boats are the real heros of the expedition and Captain Richards is unreserved in his praise of them.

THE SURVEY CREW

> I have much satisfaction however, in assuring you, that the zeal, and energy displayed by my officers, has in no way abated, and has been all I could desire, and that under circumstances of frequent, and considerable risk, I have not found them wanting in skill, or judgement. (Richards 1857–1862: June 7, 1862)

Throughout the pages of his journal and in his correspondence with Admiral Washington of the Hydrographic Office, Richards takes every opportunity to commend his officers. He particularly notes the exceptional qualities of Lieutenant Richard Charles Mayne who served with Richards from February 1857 until 1861, and who was promoted from second to first lieutenant and then to commander. Richards describes Mayne as "a rare instance of an officer, not brought up from youth in the surveying service having qualified himself by untiring applications for the highest positions he can be called on to fill." This relationship of mutual respect and admiration between Mayne and Richards is exemplified by the dedication page of Mayne's 1862 publication entitled *Four Years in British Columbia and Vancouver Island*. The dedication is made to Captain George H. Richards "Under whom I had the happiness to serve during the time I was in the colonies . . ." from "his sincere friend, the author."

Richards frequently refers to the activities of his core group of officers and surveyors. First lieutenants Mayne and Moriarty, masters and surveyors Bull and Pender, second master and ship's artist Bedwell, second masters Gowlland and Browning, master's assistant Blunden and ship's surgeon, Dr. Wood, all served with him on the *Plumper*. In January 1861,

[13] See Glossary (Appendix D) for a full description of the boats in detached service.

this same group made the changeover from the *Plumper* to the *Hecate* with two notable exceptions: Bull, the master and principal surveyor who died suddenly of unknown causes at Esquimalt just before the arrival of the *Hecate*, and First Lieutenant Moriarty who returned to England on the *Plumper*. While Bull's death was mourned and his loss felt, the departure of Moriarty went unmentioned and, a year later, when Moriarty made a request to return as senior lieutenant of the *Hecate*, Richards coolly informed Rear Admiral Washington, "I am sorry I cannot ask you to comply with his request" (Richards 1857–1862: October 7, 1861). This is the only instance in the correspondence where Richards makes what may be construed as a negative comment about any of his officers.

The arrival of the *Hecate* brought two new crew members: Hankin, who signed on as a junior lieutenant and who quickly became a valuable addition to the surveying staff, and Sulivan, a young midshipman, the son of one of Richards' old captains.

Without the admiration and respect of his crew, it would have been impossible to complete the task at hand. In his reports to Washington, Richards described the difficult conditions under which his surveyors work, and he never fails to credit them. The following excerpt is representative:

> I have now, the satisfaction to acquaint you that the whole of the western sea-board of the Island has been thoroughly examined. . . . The smallest scale on which this Coast has been plotted is one inch to the mile, much of it, on double that scale, and the harbours, and anchorages on 3 and 6 inches. To accomplish this on a coast, so entirely exposed, and in most parts, fronted by dangers, which rendered it hazardous for the ship to approach within two miles in its unexplored state, has entailed a minute examination by boats, and every portion of it, has consequently been so examined from Cape Scott, to the entrance of Fuca Strait, and I should be doing injustice to the officers, and men, employed on this trying service, were I not to bring before you the hearty and cheerful way in which they have worked under circumstances which have demanded the exercise of skill and judgement in no ordinary degree. The weather has been far more boisterous, than with a fair knowledge of the Coast I had been prepared for; the boats have constantly been detached for 10 or 14 days together, entirely on their own resources frequently unable to find any place of shelter at night, they have always returned, having accomplished more than their allotted tasks and I am thankful to say without accident. I hope therefore, you will not consider it,

> out of place, if, on the completion of this exposed coast, I mention to you especially the names of Mr Pender, the Master and Mr Gowlland, and Browning, second Masters, the most efficient officers of my staff. (Richards 1857–1862: June 7, 1862)

Richards maintained an easy discipline over his crew of over one hundred men. At a time when harsh discipline was the norm and desertion not uncommon, the journals contain no mention of flogging, and only one instance of desertion in San Francisco when four seamen jumped ship and were promptly returned. In a report to Washington sent shortly upon his arrival, he addresses the question of discipline:

> I am still under much anxiety as to the efficiency of our ships; desertion among "Satellites" Crew has become very serious and Captain Prevost has withdrawn his pinnace & crew which he had placed at my disposal. We have as yet lost no men since the great excitement [the gold rush] broke out but the temptations are very great. It would be injudicious on my part, indeed it would be fatal to our work to adopt the restrictive or prison system on board this Ship. Such a plan may have the desired effect in a vessel casually visiting the Station but stationed as we are here entirely on surveying duty, I feel that it would be totally ineffective, & I have determined to make no alteration in our system unless it becomes absolutely necessary, but rather trust to the good sense of the crew after having fully explained to them the consequences. (Richards 1857–1862: June 28, 1858)

The crew were loyal to their captain. When Admiral Maitland, newly appointed head of the Pacific Station, arrived on the H.M.S. *Bacchante*, the *Hecate* was called upon to intercept a boatful of deserters. Gowlland noted:

> The number of deserters from this ship lately has been something alarming, she has already sustained a loss of 100 men; the cause is not evident; whether undue severity in the discipline maintained on board; or from inducements held out to the men whilst they are on leave; as the cause of the running away is not as yet known; i.e. should be inclined to think a little of Each; We never lose a man; they have as much leave as they require or ask for, compatible with the rules of the Service. (Gowlland 1861–1864: August 11, 1862)

CONTACT WITH FIRST NATIONS

> In that portion of the island where the Natives are in possession you will be on your guard that no cause of offence be given, in all cases explaining by means of an interpreter the peaceable object of your visit and making the customary presents, a supply of which may be obtained at the Hudson's Bay settlement. (Washington 1857: March 16)

Today, as in Richards' time, three distinct territorial and cultural groups occupy Vancouver Island: the Coast Salish, the Kwakwaka'wakw and the Nuu-chah-nulth. Richards' survey work took him through the territories of almost every First Nation on the Island as well as to the Queen Charlotte Islands and Fort Simpson. The most detailed descriptions and personal reflections are of those groups living along the western coast of the island: the Quatsino tribes of Quatsino and Klaskino sounds, and the Nuu-chah-nulth tribes who inhabit the sounds and inlets of the western seaboard at Kyuquot, Esperanza Inlet, Nootka Sound, Hesquiat, Clayoquot Sound, Ucluelet, Barkley Sound and Alberni Inlet.

Richards was a trained and objective observer for his time, skills acquired after twenty-five years at sea in contact with cultures from all over the British Empire. He had an eye for detail that is exemplified in his descriptions of the people he encountered, like his "old friend Chief Maquinna," who arrived "dressed in a blue frock coat with 3 rows of buttons one row American Eagles the other Royal Marines. A pair of black cloth trousers and over all a long black beaver hat or Bell topper." He also provides a wealth of ethnographic detail about marriage practices, the role of women, family relationships, the acquisition of power and status, the potlatch, ownership of resources, fishing technologies, warfare, slavery, trade practices and cultural practices such as conical head shaping, a subject of great interest to him.

During the five-year period that Richards spent on Vancouver Island, both aboriginal and non-aboriginal population numbers underwent dramatic changes. In 1856, the aboriginal population of the Island was 25,873 compared to a predominantly British population of 774 living in the Victoria district, Nanaimo and the Hudson's Bay Company establishment at Fort Rupert (Douglas 1856; Akrigg 1977: 80). The colony's attempts to draw settlers had enticed only a handful of English and Scottish immigrants to settle in the rich agricultural lands in the Cowichan and Comox valleys and on Saltspring Island.

In 1858 the discovery of gold on the Fraser River brought thousands of miners to Victoria on their way to the interior. It is estimated that in 1858 as many as 30,000 gold seekers passed through the city.[14] A steady stream of First Nations, including the Haida, the Tsimshian and the Kwakwaka'wakw, were also attracted south by trade and employment opportunities, many establishing temporary villages in and around Nanaimo and Victoria. By 1860 the non-transient population of the colony had increased from 774 to 2,884, most of them living in the Victoria township (*British Colonist*, May 3, 1860).[15]

While immigrant populations increased, First Nation populations were in a state of decline. Richards frequently remarks on these signs of decline: abandoned or deserted village sites and women with no more than two children and often no children at all. This observation is supported by demographic research indicating that between 1835 and 1881, the Kwakwaka'wakw population declined by up to 80% (Galois 1994: 43).

The smallpox epidemic of 1862 had particularly devastating effects on First Nation populations. Both Richards and Gowlland gave eyewitness accounts of the epidemic at Fort Rupert:

> It is distressing to be able to afford them no assistance or relief — it could only be done by erecting a building to receive them and supplying medical assistance and comforts, neither of which are in our power, but govt ought

[14] "The excitement caused by the discovery of Gold in Frazer River is daily increasing & has already arrived at a very high pitch, 14,000 people have been imported from San Francisco up to this time & from 30 to 40,000 have taken tickets there & are coming up as fast as vessels can be got. A large Steamer is leaving that port daily, and the average number landing here now is 1,400 per week. They are not confined as hitherto to the lowest classes but many men of capital, & most of the first Merchants American as well as English are here purchasing land and establishing houses. A little more than a month ago Victoria was a quiet village. I can only compare it now to Greenwich Fair. Whole streets of Canvas Houses have sprung up. Booths Restaurants & every description of public houses have been called into existence within a few days. Hundreds of people are entirely without Shelter, and this is merely the commencement of the thing" (Richards 1857–1862: June 28, 1862).

[15] This figure did not include returns from Esquimalt, Saanich or Saltspring Island.

> to do something, funds or no funds. We leave at daylight tomorrow morning, and even if we remained have no way of alleviating their sufferings. (Richards, June 15, 1862)

The whiskey trade, too, was taking a terrible toll. Richards wrote that "it is the duty of the authorities to interpose some check, either by placing proper officials at each of the harbours, or perhaps better, by establishing native missions. The whole of the west coast of the I^d^ is allowed to take care of itself, and it is not difficult to see which way their morals are tending — as their communications with ourselves increase" (Richards, May 20, 1862). He stated his views in a letter to his commanding officer, Rear Admiral Baynes:

> It appears to me that in the present relations existing between our people and the Indian, it cannot be a matter of surprise if many wrongs are committed on both sides, and my opinion is that the Natives in most instances are the oppressed and injured parties. The white man supplies him with intoxicating spirits under the influence of which most of these uncivilized acts are committed. The white man in too many instances considers himself entitled to demand their wives or their sisters, and if such demand is disputed, to proceed to acts of violence to gain their object. (Richards 1859: August 21)

As an officer of the British Navy, Richards was nevertheless an agent of colonial expansion with a mandate to protect settlers against attack and to intervene in matters of intertribal warfare and the taking of slaves. The journal details two such interventions: one at Fort Rupert in August of 1860 and another at Fort Simpson in 1862. Richards was reluctant to use force to obtain his ends, and in both cases a "show" of force, accompanied by reason, threats and negotiations, led to a successful conclusion where no lives were lost and no communities destroyed.

Richards believed that First Nations could not be expected to live by the values and laws of the colonial government if that government took no responsibility for their education, their health and their religious instruction and that teachers and missionaries should take the place of warships and gunboats. He greatly admired the work of Anglican missionary William Duncan and the utopian community he built with the Tsimshian at Metlakatlah, though he was disparaging of the Catholic and Methodist missionaries.

Finally, he had a genuine liking and respect for the tribes he encountered:

> The only ones among them who seem to have the slightest idea of dishonesty are those who have visited Victoria and become acquainted with European customs. At the same time it is only fair to say that it is scarcely right to judge by what happens to us. They naturally dread our force, and may be deterred from taking advantage from this cause. Still, our boats are constantly among them far removed from the ship and quite in their power. They have been always foremost to help them, landing in a Surf a whole village has come down to haul the boat up or to launch her, and they have always shewn the most friendly feeling. I can safely say, having seen & had dealings with almost all Native tribes in the world. I have never met a more friendly, harmless and well disposed set of people than those on Vancouver Island. (Richards, May 20, 1862)

— Linda Dorricott & Deidre Cullon
Nanaimo, 2011

Note on the Text

Captain G.H. Richards and Thomas Gowlland both have easy, readable styles; their narratives flow naturally and many events are described with literary flair. Every attempt has been made to reproduce the journals with as few editorial changes as possible.

The original journals contain numerous spelling errors. These errors have not been noted to avoid peppering the text with [sic]. First Nation names have been retained without correction. An index of First Nation names and their spelling variants is attached as an appendix. Abbreviations for time of day, measurements, readings, etc., remain as they appear in the original.

No attempt has been made to standardize ships' names; both Richards and Gowlland occasionally put ships' names in double quotation marks but they are usually written as proper nouns, and usually without quotation marks.

Both Richards and Gowlland capitalize many words that today would not be considered proper nouns. Because it is not always clear whether the frequent use of a capital letter is common usage for the time or idiosyncrasies of handwriting, in most cases they have been reproduced as they appear in the handwritten text.

Underlined words, question marks, cross-throughs and asterisks appear in the text as Richards and Gowlland wrote them.

Some aspects of the text have been standardized for the sake of clarity. Date entries have been standardized. A typical date such as "Tuesday 11th Septr has been rewritten as "September 11, Tuesday." Days of the week are included only if noted in the text.

Richards' punctuation is inconsistent in the first year of the journal, but by 1861, as the journal entries become longer and more descriptive, these inconsistencies give way to more standardized punctuation. Richards and Gowlland both make frequent use of dashes and commas. These have been retained as long as the meaning of the text is clear. If the text is ambiguous and there is no clear indication of the beginning or end of a sentence, a period has been added along with a new sentence beginning with a capital. A small number of dashes and commas have been removed.

Square brackets denote an editorial correction and are used when the present day name is different, (e.g., Admiralty Island [Saltspring Island]) and when the present day spelling for place names is different, (e.g., Ballinac [Ballenas]). For words that are indecipherable, the word "illegible" is enclosed in square brackets, i.e. [illegible]. Richards occasionally left blank spaces in the journal which he presumably intended to fill in at a later date. These blank spaces are indicated by horizontal lines.

Excerpts from Gowlland's journals are set off from Richards' journal by double indents and horizontal lines before and after. When dates do not exactly match, the excerpts are placed so that the flow of the narrative is not interrupted. Gowlland kept three journals and excerpts from each of these is included. The dates at the end of an entry indicate its journal of origin. Editorial excerpts are indicated with ellipses [. . .].

Gwa'sala-Nakwaxda'xw
Tsawataineuk
Tlatlasikwala
Gwawaenuk
Da'naxda'xw-Awaetlala
Kwakiutl
Mamalilikulla Qwe'Qwa'Sot'Em
Fort Rupert
Kwicksutaineuk
Quatsino
Alert Bay
Tlowitsis-Ma'amtagila
'Namgis
Laich-Kwil-Tach
Ka:'yu:'k't'h/Che:k:tles7et'h'
Cape Mudge
Ehattesaht
Nuchatlaht
K'ómoks
Comox
Mowachaht/Muchalaht
Gold River
Yuquot
Hesquiaht
Ahousaht
Tseshaht
Nanaimo
Snuneymuxw
Tofino
Toquaht
Ucluelet
Uchucklesaht
Ucluelet
Huu-ay-aht
Cowichan
Esquimalt
Victoria
Songhees
N

Vancouver Island First Nations

▲ Vancouver Island North

Port Harvey
Port Neville
Johnstone Strait
Bute Inlet
Chatham Pt
Seymour Narrows
Discovery Ch
Quathiaski Cove
Campbell R
Cape Mudge
Tahsis Inlet
Nootka Island
Gold River
Muchalat Inlet
Yuquot

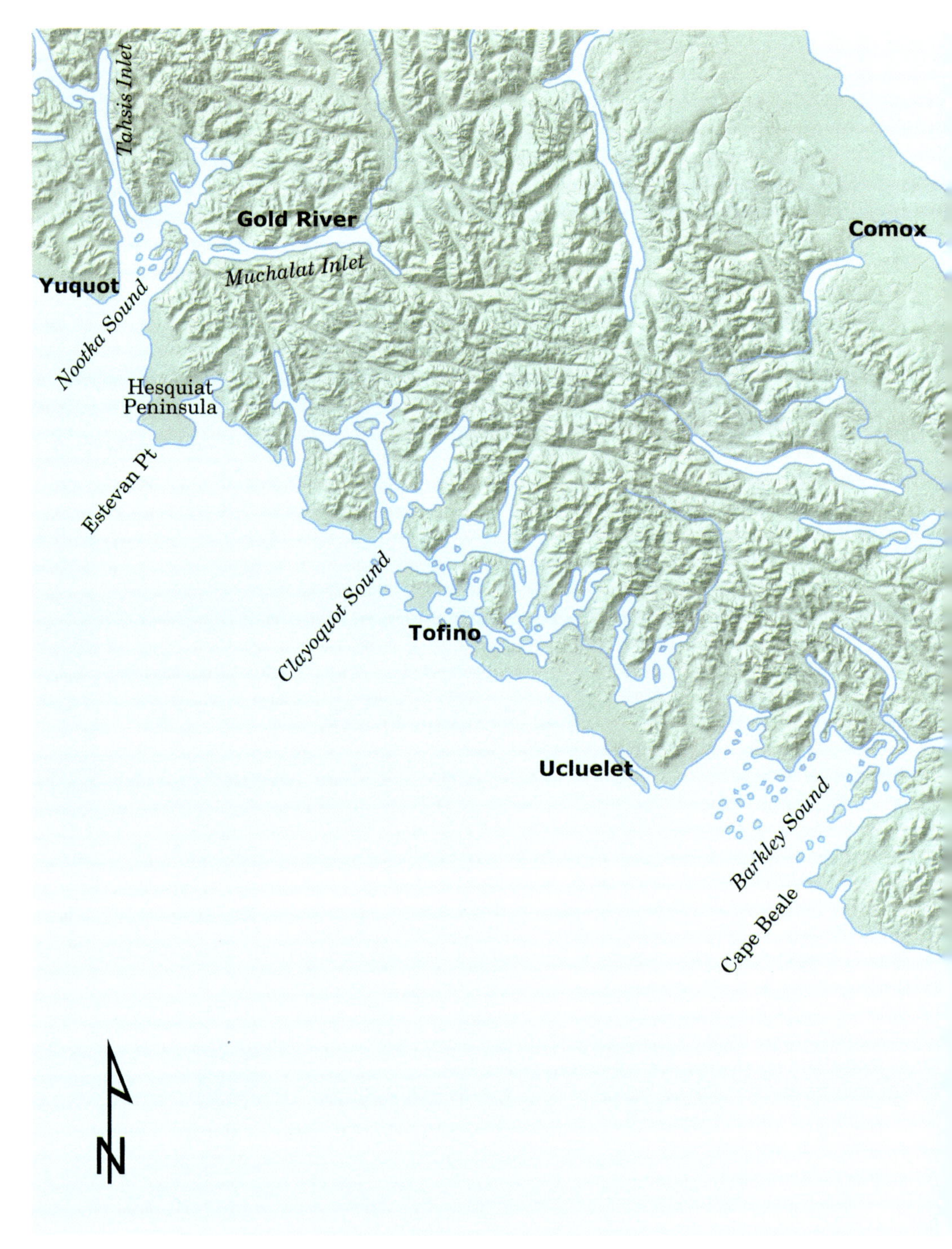

▲ Vancouver Island South

Jervis Inlet
Malaspina Strait
Texada Is
Cape Lazo
Denman Is
Hornby Is
Baynes Sd
Lasqueti Is
Strait of Georgia
Howe Sound
Burrard Inlet
Port Alberni
Nanaimo
Gabriola Is
Fraser R
Alberni Inlet
Point Roberts
Galiano Is
Tincomali Chan
Active Pass
Admiral Is.
(Saltspring Is.)
Cowichan
Haro Strait
Bonilla Pt
Port San Juan
Juan de Fuca Strait
Cape
Flattery
Neah Bay
Esquimalt
Victoria
Race Rocks

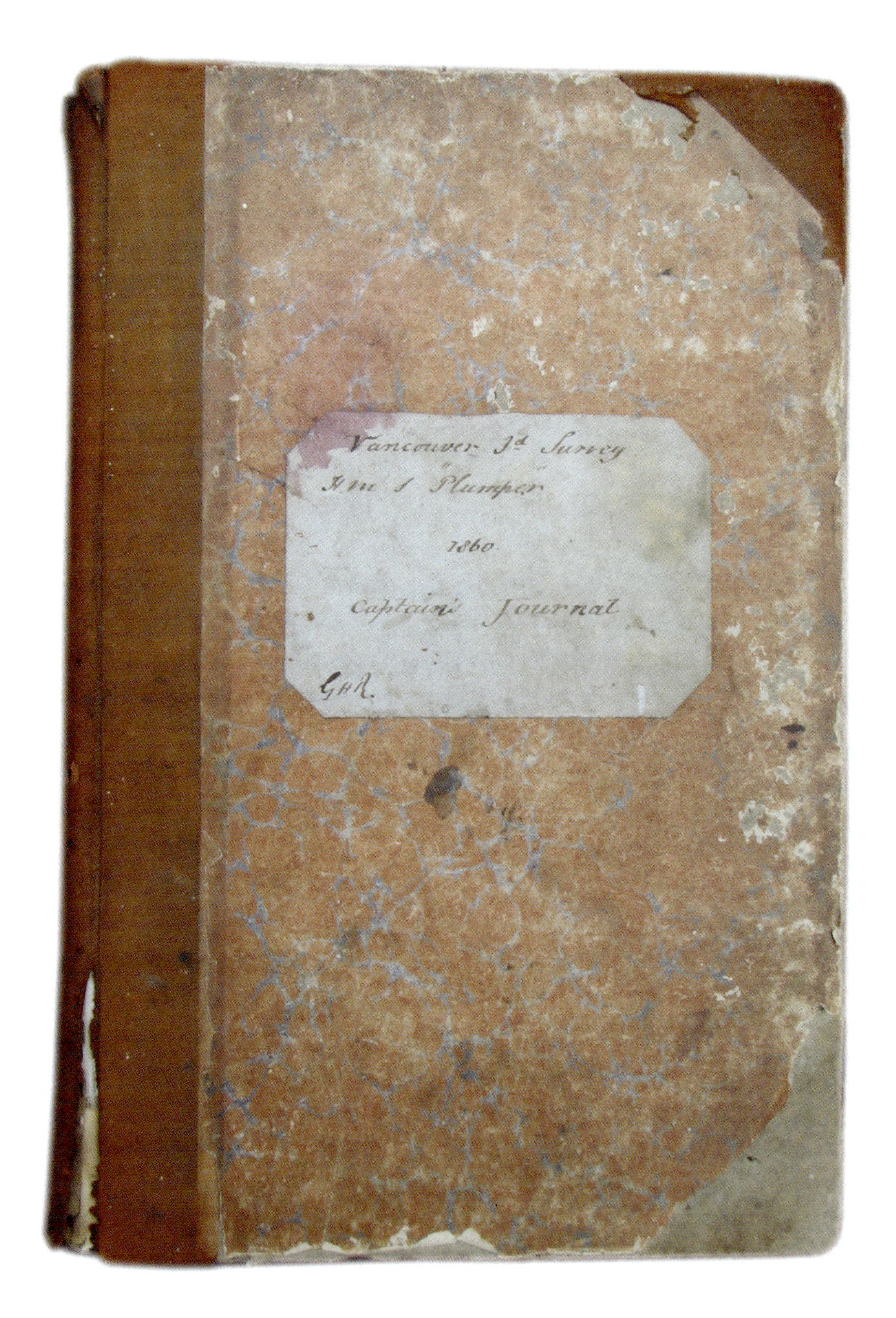

▲ Cover of Captain Richards' journal. (PHOTO: DEIDRE CULLON)

▲ A page of Captain Richards' journal. (PHOTO: DEIDRE CULLON)

▲ Donal Channer in his library. (PHOTO: DEIDRE CULLON)

§

Vancouver I^{d} Survey
H.M.S. "Plumper"
Captain's Journal
G.H.R.

1860
Survey Season

ESQUIMALT

The winter of 1859–60 has been a remarkably fine one — comparatively but little rain has fallen and no heavy gales have occurred — the winds have been principally from the Northward. NNW to NNE — but not an instance of a strong NW wind has occurred during the winter; two intervals of extreme cold have taken place — lasting each for three or four days — during these snow lay on the ground from 1 foot to 18 inches in depth.

ESQUIMALT TO SAN FRANCISCO

FEBRUARY 10. Left Esquimalt for San Francisco, weather unsettled and Barometer 29.90 falling — found a strong SW^{r} at the entrance of Fuca Strait and anchored at 3 PM in Neah Bay.

ESQUIMALT

APRIL 1, SUNDAY. Ready for Sea.

APRIL 2. A SW gale prevented our starting today — the Baro. fell from 29.90 to 29.75 and it blew a fresh gale all day, unmoored in the morning — but were obliged to drop a 2d anchor before noon. The Georgiana a coal trader from Nanaimo for S. Francisco came in — were put back.

APRIL 3. Weighed one anchor at daylight — but the weather would not allow us to start. Baro fell during the day from 29.70 to 29.22 and every sign of a south east wind — hail and snow fell in the afternoon, before midnt — we let go our 2d anchor again; after midnt it blew Violently from the Eastward the barometer still falling.

APRIL 4. At 4 am a very Violent gale — trees being blown down in great numbers — baro at 8 am 28.93 the lowest known here since our arrival at 8.30. Commenced to moderate. Baro rising and Wind veered to SW blowing fresh all day, moderated at eve — glass rose rapidly up to 29.53 then stopped.

❧ ADMIRAL ISLAND

APRIL 5. Calm this morning — but heavy cloud to SE. Glass only up to 29.53. At 7 weighed and steamed out — intending to anchor on and examine the Salmon bank off South end of San Juan which the Satellite[1] is supposed to have struck on but which I don't believe — passed Satellite off Trial Ids — bound for Esquimalt — a fresh SE wind springing up — gave up Salmon bank and Steamed up Haro Strait against the tide. Hail and Snow Squalls all day. Passed up Trincomalee Channel — thro Ganges Harbr and anch in 9½ fms — off the E side of Admiral Id [Saltspring Island] 3 miles south of Pt Southy ⅓ of a mile from the shore — with shoals rather suddenly inside us. Landed to enquire into a complaint made by the settlers of Salt Spring Settlement regarding the natives. Alpha — Coal Schooner called here from Nanaimo to Victoria at 4 PM. There are about 20 settlers here English and Canadian. 6 or 8

---

[1] Since 1856, the H.M.S. *Satellite* had been under the command of Captain James C. Prevost, the first British Boundary Commissioner appointed to resolve the dispute over the British-American maritime boundary. In 1853 Prevost charted the waters of eastern Vancouver Island and Haida Gwaii [Queen Charlotte Islands] on the H.M.S. *Virago*. Richards became the second Boundary Commissioner in 1857 (Gough, 1974: 154–55; Mayne, 1862: 10).

houses. Each house with 3 to 5 acres of land fenced and under cultivation. The soil appears a rich dark loam about 2 feet thick then clay. Saw the salt springs — they are not above 100 yards from the beach.

The Settlers have no complaint against the Natives — they merely wish to keep them off the island — which seems scarcely reasonable as the Indians have used it as a hunting ground from time immemorial and have received no compensation from Government prior to settlers being established. The indians always like to have the whites near them for obvious reasons, they get paid for their labour and derive many advantages — but if they are excluded from the white mens settlements these advantages will cease and the Europeans will become inimical to the native.[2]

APRIL 6. Weighed at 7 AM and ran up to Dodd Narrows — full moon yesterday — low water at 10.45 today — passed up ½ an hour before low water tide from 2 to 3 knots ebb against us — the high water at these narrows appear to be at 4.30 AM on F&C[3] days — during all the year — dropped the gig to seek for anchorage between Dodd and False narrows and steamed on for Nanaimo — anchoring at 11 AM, no coal to be got today — in consequence of the low tide; several schooners here loading from Victoria and Fraser River.

APRIL 7. Got observations for rates of chronometer and latitude — SEly wind. Cold and weather unsettled. Gig returned having found anchorage between False and Dodd Narrows which I called Percy Anchorage.

---

[2] In July 1859, a "Land Reform Meeting" was held in Victoria setting the stage for settlers to move to Salt Spring Island and Chemainus. By fall of that year, approximately thirty people were eligible for land on the island (*British Colonist* 1859–1862: August 15, 1859 and September 21, 1859). In March 1860, the House of Assembly recognized the need "to provide means for extinguishing, by purchase, the native title to the Lands in the districts of Cowichan, Chemainos and Salt Spring Island, which are now thrown open for settlement. The purchase should be effected without delay, as the Indians may otherwise regard the settlers as tresspassers and become troublesome" (*British Colonist* 1859–1862: March 3, 1860). Mayne echoes Richards' sentiment: "The Indians here have given no trouble, only occasionally informing them . . . [that] the whites had no business there, and that the land belonged to themselves. . . . They appear anxious to see Mr. Douglas & to be paid for their land" (Mayne 1857–1860: April 6, 1860).

[3] Full and Change. See Glossary for definition.

It is very convenient for vessels going to the South^{d} through the narrows as they can wait there for the tide — sent letters to Admiral by the Osprey Schooner.

APRIL 8, SUNDAY. Fine day got observations as yesterday. Wind fresh from West & NW.

NANAIMO AND KOMAX [COMOX]

APRIL 9. Wind ESEly. Gloomy morning. Baro 29.20. Sent Shark to sound on West side of Gabriola I^{d} and Fairway channel — left at 8 in galley to observe angles on Entrance and Lt H^{s} I^{ds}. Strong breeze set in from ESE. Ran back by Departure Bay. Landed at the mines on Newcastle I^{d} which are now being worked — the Coal apparently very superior.[4] Wind freshened to a gale with heavy rain and continued so all day. Baro however only fell 2 tenths; bought a whale boat off 2 Americans for 165 dollars for surveying work. Got in about 20 tons of coal today by a Skow. English barque "Sea Nymph" arrived for coals.

APRIL 10. April Wind Westerly or NWly fresh breeze and fine w^{r} apparently blowing strong in the Strait of Georgia — as it generally does — at this place with NW winds. Completed Coaling by a lighter — sent tracings to Hydrographer correcting an error in the River off this place — in charge of Cap. Franklin [Franklyn] who goes down in a canoe.

APRIL 11. Fine morning but cold. The season seems 14 days later than last year, and not settled for boat work. Wind S.W or [illegible] observed for Variation of the compass — and Star latitudes.

APRIL 12. Gloomy in the morning but cleared and proved a very fine day. Got Equal altitude for rates and at 2.30 PM left for Qualicom river 34 miles to the NW. Passed between Five fingers I^{d} and the rocks to the west^{d} of them — a good deep Channel for Steamers. Steered thro Bal-

---

[4] The first shipment of coal left Nanaimo in 1852 and, by 1853, production was smooth and the quality of coal was considered good. By 1862, the Douglas Pit was operational and the coal was considered "superior to anything yet found in Nanaimo and better for steaming than coal from Sydney, Cape Breton or Pictou, Nova Scotia. In fact, it was equal to most English coal for steaming; it stowed better than Welsh coal, and . . . there was lots of it" (Bowen 1987: 126).

linac [Ballenas] Channel and at 8 PM anchored off Qualicom river in 9 fms ½ mile from shore. It has been calm all day in the Strait and the mountains more clear than I have ever seen them.

HENRY BAY BAYNES SOUND

APRIL 13. Weighed at 4.30 AM in order to reach Henry Bay at the North end of Denman I^{d} in time for observations — dropped the whaler with 6 days provisions off Hornby I^{d} and proceeded thro Baynes Sound anchoring at Henry bay in 13 fms. at 8 AM. The weather rather gloomy but got observations for the Meridian distance — Shark and Gig went sounding. The natives from Port Augusta [Comox] Village Visited us immediately — friends of last year.[5] Deer geese and fish plentiful enough.

APRIL 14. Repeated observations for time and latitude. Got the Dep and Variation. Shot 3½ brace of humming birds and one very handsome Woodpecker. Fired numerous round of Minie rifle ammunition[6] at a Bald headed eagle without effect and commenced to cut down a huge tree in which the nest was built, hauled the seine and caught quantities of herring. A party of Indians came to our observation point with their worldly goods — consisting of mats & cedar planks roughly hewn with their stone axes — wherewith to construct their habitations which they did before nightfall, harmless dirty creatures enough. The boys shooting small birds with bow and arrow. Women louzing each other. Fish and dried clams their food. The latter esteemed a delicacy by some of our officers — they scewer them on a stick and dry them in the sun — it is to procure this delicacy that they visit this point and if one may judge by the quantities of shells of which this end of the I^{d} is almost composed to the depth of several feet, their ancestors must have resorted to the same spot for generations past for the same purpose. The clam Shell makes excellent lime by merely burning them from 24–48 hours.[7] Digging a

[5] For a list of First Nations met by Richards and his crew, as well as their current names, see the Index of First Nations.

[6] The Minié was a nineteenth-century rifle designed to allow rapid muzzle loading.

[7] Lime was used as whitewash for painting survey marks on the shore. Lime was made from shells by heating them for a couple of days in a "kiln of stones and fine bars." According to Gowlland, it was "as good as the limestone for . . . whitewashing marks" (Gowlland 1861–1864).

large hole and placing alternate layers of wood and Shells. When sufficiently burned cover the whole thickly with seaweed & let it remain 12 hours. Then serve — and head up in Casks. The Northern Entrance of Baynes Sound — from this pt to Cape Lazo is a width of nearly 5 miles but is very shallow. A bar of sand actually right across between the two points. There is not above 14 feet on this bridge at low water and when we crossed it last summer it was thickly covered by kelp. Now there is not a particle of kelp growing on it. It is very remarkable that the kelp disappears always at the end of summer — certainly always in exposed situations. Whether it dies or is dispersed by winter gales I know not. I suspect the former but at any rate it must grow with remarkable rapidity for in two months time the bridge will again be covered and the long Stems of kelp 6 or 8 fathoms in length. A party of Fort Rupert Indians in 2 Canoes passed this Eve for Victoria.[8]

KOMAX–DENMAN I^D^

APRIL 15, SUNDAY. A very fine day but somewhat chilly — the near range of hills opposite over the Vancouver Shore[9] 6 miles inland and from 2 to 4000 feet high are thickly Covered with snow. The land between the sea Coast and these hills is low and thickly covered with trees — which are the only impediments to agricultural undertakings. Prepared 2 boats to proceed in advance to Cape Mudge — examine the Coast between this place and it and find anchorage for our next astronomical station. The Ship must remain here until Thursday to secure rates etc.

APRIL 16, MONDAY. At 5 this morning Gig and Shark left for Cape Mudge — and Lt Mayne & Dr Wood started shortly after with Canoe to explore the Courtenay River which flows into Port Augusta.

Weather fine but cold and rather gloomy. Wind Northerly. Shark did not round Cape Lazo until after noon. Baro falling and below 30.00. The Indians from the Various Villages are collecting at this point — and

[8] Victoria was the trading hub of the Colony in the early 1860s. Many coastal First Nations travelled by canoe to Victoria to obtain better prices for their furs.

[9] Richards frequently refers to the mainland of British Columbia as the "Vancouver Shore."

bringing the moveable parts of the Villages with them. We have had an accession of about 40 today and now number about 80 here. They appear to shift from one Village to another according to the season. In Summer they generally encamp at the sheltered head of some bay — for convenience of gathering berries where they remain till Spring. Then as at this time they shift to the Island bays and sea coasts for fish and clams. I observe but few young people among them. The few there are are better dressed than the old ones, the women having smart dresses and painted faces. They are generally extremely dirty yet they are anxious to get Soap and while we are here some of them used it — particularly the few young people and children. The whaler returned this evening having completed the work she was left to do on the 13th at Hornby Id.

APRIL 17. Got observations for time and lat. In the afternoon a canoe arrived from Nanaimo bringing our mail — letters to 17 Feb. Heard that Hecate was to [be] fitted and sent to us.[10] Landed for Stars tonight but weather too cloudy to get them Satisfactorily.

APRIL 18. Rain and strong SE wind all day. Confined to ship. Canoe could not start back with return blanket letters. Mayne & Wood returned last eve having explored the land about the Courtenay. There does not appear to be such an extent as we believed. Mr Mayne describes about 7000 acres of excellent grassland — river navigable for boats for about 2 miles — then divides — one branch running NWd thro the low land and the other to the westd between Mt Becher and the range known as the [illegible] — the latter branch the indians call the Puntluch _______ the former _______ or the Courtenay of our chart. I think it very probable that much more good land exists than the Indians were willing to show. They evinced a great indisposition to go at all — from their laziness I believe as they have at present deserted the Village at the river's mouth for the sea Coast and moreover our party was only away two days from the ship and the weather unfavourable for exploring.

APRIL 19, THURSDAY. Cold SE wind — unable to move today as I had intended; the bad weather of yesterday preventing our completing some

[10] Richards heard, unofficially, from his brother, that the navy was sending the *Hecate* to replace the *Plumper*.

soundings and getting satisfactory observations for lat and time. Landed this morning for them. Succeeded but partially. At 9 am a canoe was seen approaching from the South[d] with 3 flaunting banners of bright yellow. The Natives recognized them as the priests and on a volley being fired from the Canoe a commotion ensued at the Village now numbering from 100 to 150 Indians.[11] They have almost ousted me from my Observation spot by erecting their houses right over it. The holy fathers landed on the beach at 9.30 the greater part of the Indians being ready to receive them. They were dressed in petticoats — one with a bell topper — the top [illegible]. The other fellow in a black Jim Crow — appeared much more comfortable. They harangued the multitude for a few minutes and then the yellow banners were removed to a convenient spot on shore where the natives immediately commenced a house of planks & mats either for a Church or temporal residence for the shepherds. I was glad to see the tabernacle erected some 300 yards from me for the sake of my observations. There appeared a few exceptions to the attendance. One old chief who was on board the ship steadfastly persisted on remaining altho no attempt had been made to convert him to Protestantism by us. The old women also appeared to be excluded from the tabernacle. The[y] pursued their ordinary avocations — as usual without in the slightest degree noticing what was going on. Launched their Canoes, dry Clams and prepared the food. Got Lat and AM & PM sights. Conversed with the priests who are on a journey of inspection preparatory to establishing Missions. They travel about in a miserable way — a small canoe, feeding on Native fare, preaching their way in fact instead of paying it. The Village has increased wonderfully, a week since not a man here, now nearly 300 people with 61 considerable large Canoes and the point covered with their houses.[12] Got star latitudes this evening. If

---

[11] By 1858 a number of Roman Catholic priests and missionaries were based in Esquimalt. These included Reverend L.J. D'Herbomez; Fathers Ricard, Chirouse and Pandosy; Sub-Deacon Brother Blanchet and Janin, a lay brother (Theodore 1939: 228–230). By 1860, Roman Catholic, Anglican and Methodist missions were competing for converts among the First Nations of Vancouver Island.

[12] Around 1846, the K'ómox people were forced southward by the Laich-Kwil-Tach First Nation, leaving their homes around Quadra Island and moving in with the Pentlatch at Comox. The Comox area on Vancouver Island and the Strait of Georgia were well known among local First Nations as an area rich in resources. The word *K'ómox* is from

howling is any evidence of piety then Indians are on the right road. They kept it up till nearly midnight — it was intended as Sacred music. The tides appear more regular here — I should say the regular 6 hour interval — high water at the F&C about 6 o'clock PM.

§ DENMAN ISLAND AND CAPE MUDGE, DISCOVERY CHANNEL

APRIL 20, FRIDAY. Weighed at 5 am. Holding ground so good that we had difficulty in getting our anchor. Steamed across the bridge which Connects Cape Lazo with the North end of Denman I[d] at a little after high water — 23 feet the least water — met a strong S.E. wind outside and found an extensive and dangerous bank extending SE of Cape Lazo for 2 miles — after sounding it Steered for Cape Mudge on the high land of Valdes I[d] [Quadra Island] behind it. The Cape is a cliffy one about 100 feet high or rather a range of cliffs similar to Cape Lazo. A bank also extends off it for a considerable distance to the South[d] consisting of large bolders. M[r] Pender joined in Gig at 10 am having found anchorage for the Ship 3 miles within Discovery Channel on West Side of Valdes I[d] which we proceeded for; a considerable sea caused by the strong SE wind; all the coast between Cape Lazo and Mudge is composed of large bolder stones on which it is dangerous to beach boats — & much exposed the land rises to a moderate height wooded and there is a very considerable quantity of comparatively low land between the Coast and the Mountain ranges tho probably not cleared; we found a very strong flood tide running from the Northward and with sail and steam made but little progress. Heavy tide rips exist at the Entrance of Discovery Channel — the flow from the North meets that from the South[d] about Cape Lazo as we supposed last year. At 1 PM we moored in the small

---

the Kwak'wala word *k'umalha* meaning "plenty." The waters around Denman Island teemed with herring each spring, seals were trapped at the north end of Denman Island at Seal Islet and the tidal flats at the mouth of the Courtenay River were covered in dozens of large, complex fish traps. Every year the Laich-Kwil-Tach joined the K'ómox here. They travelled with their house planks strapped between two canoes, making a catamaran, and all goods were transported on this deck. Upon arriving at the temporary village or campsite, the planks were used to quickly erect a house which would be dismantled again at the time of departure. This system was used by almost all First Nations of Vancouver Island.

bay recommended by Mr Pender — Quathiosky Cove and found the Shark here. There is only room for one vessel to lie comfortably in this small place. 2 vessels could find shelter and there is room for several small vessels within Northern arm. It appears a good stopping place, is out of the tide and perfectly secure. Got afternoon obs^{t} on a small islet at the north pt of the Entrance. The night fine as is generally the case — indeed a clear starlight night — appears to be the precursor generally of a rainy day — got some good star latitudes.

APRIL 21. A very gloomy dull morning — no chance of observations. A native or two came onb^{d} bringing a live Beaver — for which they wanted a Blanket — boats away surveying. As I could not get observations went away & marked the Coast on Valdes I^{d} Shore for 4 or 5 miles preparatory to the Survey Northwards — found a very good harbour 2 miles north^{d} of this Cove — it will always be recognized by an Island[13] (wooded) lying off it. Pass in North^{d} of the I^{d} & then Steer up the bay to the South^{d} — where excellent anchorage is had in 9 fathoms. It rained all day and towards night came on more constant. Indeed the weather is quite wintry and unlike any season we have experienced here. Hitherto the place reminds me very much of the Straits of Magellan — all boats returned by 5 PM — prepared them for detached work on Monday Morning.

> APRIL 20. On Valdez Island 2 miles inside the Cape are the remains of an old Indian Village which from up appears like a stockade at present uninhabited. . . .[14] The name of the Tribe that inhabit this part of the country is Eukaltah — they are a

[13] Gowlland Island.

[14] This is the site of the current Cape Mudge Village and Indian Reserve on the western shore of Quadra Island. "Eukaltah" or Laich-Kwil-Tach people are the southernmost Kwakwaka'wakw group. By the mid-nineteenth century they had expanded southward and were centred at Cape Mudge and Campbell River. They were well known on the coast for their warlike behaviour. Mayne referred to them as "the Ishmaelites of the coast" (Mayne 1862: 245). In the summer of 1860 the Cape Mudge village was bombarded by the gunboat H.M.S. *Forward*. The Laich-Kwil-Tach retaliated with muskets, and the skirmish ended with canoes destroyed and the stockade in ruins (Gough 1984: 132; Mayne 1862: 245–246).

> very warlike savage race; always fighting with their neighbours and mostly gain the advantage — tis said that Canablism is not extinct with them yet; but I very much doubt the truth of the report, as they appear much too fond of deer and Salmon to wish for any more delicate food. . . . The Indians keep the Ship's Company well supplied with deer, Salmon, and Wild-fowl. The former we traffic for Shirts & latter Biscuit, tobacco or Soap, one large Elk I brought away in the pinnace could not have weighed less than 3 cwt — for an old coat. Powder flask, price of Tobacco; soap and a few beads in value altogether about 5/- [5 shillings] (Gowlland 1859–1864).

APRIL 22, SUNDAY. Tho the Barometer was very high today up to 30.30 yet the weather is most uncertain. One hour exceedingly fine & clear — then cloudy. Cold and wintry Wind SWly — got some indifferent equal Altitudes — 3 or 4 Small canoes of very miserable indians attracted to the cove by the Ship's presence probably. They brought deer grouse and a very good kind of rock fish. The Romish priests hove in sight at 3 PM and came into the cove with colours flying; just as we were sending to offer civilities to our faithful Allies they set sail and steered for the bay 2 miles to the North[d]. I fear they do not look upon us as Allies. All boats ready for an early start tomorrow morning.

APRIL 23, MONDAY. Boats away at 5 o'Clock. Morning fine. Baro 30.40. Went down to Cape Mudge against a Strong ebb — erected a great Mark there — and returned — finishing the East side of the Channel on my way; a large Village a mile within the Cape but no Indians there at present. Several fellows triced up to the trees in boxes — with their blankets hanging over them — reminded me of the Table Hatchment which figures over the residence of a more civilized Savage in Belgrave Square — why is not a blanket as good and appropriate.[15] These fellows — Ughcultas were murdered by a party of Hyders a year ago — on their return from Victoria — saw Mount Conuma today — our Nootka

---

[15] On the Northwest Coast, it was a common funerary practice to place the deceased in a spruce tree.

Sound in Vancouver's Chart; I think the strength of the tides in this Entrance is from 3 to 5 Knots.

APRIL 24, TUESDAY. Morning very fine. Mountains very clear — fine distant ranges. I think from 5000 to 8000 feet very marked in their rugged outline and covered with Snow. Got Sketch of them. Started early and finished the Strait as high as Menzies Harb^{r}. The distance is not so great as represented in Vancouver's chart by 1 or 2 miles; besides the Cove[16] we are in there is a good large harb^{r}[17] 2 miles above us on this Shore and anchorage on the Western side of the Strait in one or two places out of the tide; a considerable river[18] empties itself into the Strait on the Western & Vancouver shore — abreast our present anchorage — with some clear land on its banks — no cedar trees are found in our neighbourhood at least but very few — and they particularly small; very few natives — a family or two are encamped near us. I have just been driven off deck by the children squalling — the night being still — ship 150 yards from the bank we hear them plainly — they remind me of my own. I can't detect any difference in the notes. 2^{d} Whaler Mr Bull returned tonight from outside.

APRIL 25, WEDNESDAY. Morning cloudy. Started 6.30 am to survey Menzies bay and connect ships work with that of the boats gone in advance — rained nearly all day. We have certainly a different climate to the Southern end of the I^{d}; I suspect being among the Mountains and confined by narrow channels makes a difference in our weather which the Barometer does not indicate.

Menzies bay immediately South^{d} of the Narrows affords good anchorage. There is a dry bank of mud at low water in the middle of the head of the bay; within which there is good anchorage in 7 fms — it is best to pass in the North side of the bay keeping the north shore onb^{d} within 100 yards.

The head of the bay is steep — 6 fms close in at low water — and no tide is felt; the flood setting from the North down the Narrows sets across to

16 Quathiaski Cove.

17 Gowlland Harbour.

18 Campbell River.

the South Shore of Menzies bay with great strength — 8 or 9 kts — but there appears no difficulty in navigating with a Steamer — it is only necessary to attend to the tides and anchorages are abundant — the tides do not turn with such great rapidity as I have seen in some reports for half an hour or more — there was no tide to speak of between the conclusion Ebb and Commencement of flood. Finished Menzies Bay and connected our work with the Narrows. Returned to ship at 5 PM. Shark returned from Cape Mudge at 9 PM.

❦ QUATHIOSKY COVE DISCOVERY CHANNEL

APRIL 26. Gloomy rainy day — Shark left with 10 days provisions for the North[d]. Landed in the cove to look at a remarkable seam of quartz about 12 feet wide cropping up thro the trap rock. About a dozen Indians camped in the cove — chiefly attracted by the Ship — they are of the Komoux tribe & miserably clad. They supply us with an exceedingly good kind of fish a cod I think which is caught with hook & line about the rocks. The red fish[19] is also caught here, grouse are pretty numerous likewise deer which our people patronize regardless of the Season.

APRIL 27. Gloomy morning. Strong NW wind Baro down to 29 80 — 2[d] Whaler went down to Sand Bank off Cape Mudge — did not get back till 8 PM., 1[t] Whaler with Pender returned at 4 pm. from Port Neville 50 miles West[d] of us. He reports it a good harb[r] — he met Shark and Gig — at work above the Narrows today — blew very Strong all day but got Equal Altitudes — got off Tide party and prepared for a start tomorrow morning; the native name of this cove is Quathiosky.

❦ KNOX BAY DISCOVERY CHANNEL

APRIL 28, SATURDAY. A strong SWly wind — unmoored and started shortly after 7 am and steamed against the last of the flood for Seymour narrows dist 8 or 9 miles. Wind blowing in strong gusts. Ship making but little way — shortly after 9 passed the Narrows which are about half a mile in width opening out to a mile shortly after passing thro. The Ebb has just made and ran perhaps 4 or 5 kts — with the Strong wind against

---

[19] Red snapper (*Sebastes ruberrimus*).

us we made but slow progress, the reach runs nearly true north almost to P^{t} Chatham, mountains rising abruptly from the Water — to 3000 feet — on the Vancouver shore — and almost to the same altitude on Valdes I^{d} side, passed P^{t} Chatham at 12.30. It is about 12 miles from the Narrows. A rock stands about a Cable off it — or rather North^{d} of it — which is scarcely in the way of Ships. Wind increased to violent squalls and blew right down the reach which after rounding Chatham pt. turns to NW^{d}. At 1.30 pm the tide began to slack and wind blowing very strong we put in to Knox Bay anchoring in 13 fms about 70 yards from the Shore off the Mouth of a Stream, the water is very deep in this bay — until close in — 60 fms then 30 and 14 — it is however a good stopping place for a steamer which can pick her berth up easily — but one insecurely deep for a Sailing vessel — got observations, at the stream for true Bearing — & measured a base — to commence a Survey from. At Night — got Latitudes by Stars.[20]

In the evening 20 canoes of Hyders from Queen Charlotte I^{d} came alongside en route to Victoria — they are a wild independent — set of fellows — very superior to the Tribes farther South — physically and intellectually. They had Skins on b^{d} and were accompanied by their women. I saw no children or very few. They ask a larger price for their skins than they could be bought dressed in Regent Street for — and our people unaccustomed to trading and eager to get any thing spoil the market — they left us reluctantly at dusk camping close to and starting at 4.30 am with the tide on their journey.

DISCOVERY STRAIT–PORT NEVILLE

APRIL 29, SUNDAY. Weighed at 8.30 am for Port Neville this not being a desirable spot for a sojourn. Left 2^{d} Whaler with Mr Bedwell to take up

---

[20] Richards' surveying tool kit was likely similar to that of a land surveyor of the time. Essential survey equipment in the 1860s would have included a plumb-line, a level, levelling rods to measure elevation, a chronometer to measure time and longitude, logarithm tables to calculate trigonometric functions of bearings and a compass to establish magnetic directions (Gordon 2006: 55–56). Richards later mentions his theodolite, a device used to measure horizontal and vertical angles. To survey accurately the shoreline and to chart the waters, Richards and his crew established regular measured bases. All other measurements were established by triangulating to this point, resulting in accurate survey measurements.

the Survey tomorrow and bring it in to Helmken I^{d} 12 miles to the Westward — had the first of the Ebb with us — at 8.30 am. The Mountains rise abruptly from the sea on the Vancouver shore immediately after leaving Knox Bay to 6000 or 7000 ft covered with Snow. Saw some magnificent Snow Mountains on the main land side. Also, the depth of Water is very great. We only reached bottom with 140 fms of line not ½ a mile from the Shore. We are now left with 2 boats the Galley & the 1^{t} whaler; Shark, Gig and 2^{d} Whaler being detached. Passed the Salmon river just westward of Helmken I^{d} at noon — this appears a likely place for anchorage* [Richards' footnote: "no anchorage in Salmon River"] and there appears to be a pass across the I^{d} from here, a valley taking the direction of Nootka and another in the direction of Clayoquot. Entered Port Neville at 1 PM — a small I^{d} lies off the northern entrance pt. The entrance of the port is narrow and has no more than 23 feet at lower water — it continues narrow for ½ a mile or more and we got as little as 18 feet — it then opens out and runs to the East^{wd} forming a very spacious and excellent harbour with a depth of from 6 to 7 fms. Anchored on the North Shore 2½ miles within the Entrance close to a very peculiar Nob[21] of land — projecting from the shore and joined to it by a low strip. On this I pitched my tent and made it my observation spot.

APRIL 30, MONDAY. A Strong Westerly wind blew all day & prevented our sending a party outside — got observations myself for time, Lat and True Bearing and measured a base preparatory to the Survey. 6 Canoes with Matalpie Indians came in at 5 PM and landed on our observation Nob — well behaved people — with apparently but little to eat but dried herrings — and came in I fancy — with the idea of bettering their condition thro us — having seen us coming in from outside yesterday.

MAY 1, TUESDAY. Scarcely summer yet — Strong Westly wind all day — and very cold — Bull with Whaler left for a week to survey towards Helmken I^{d} this morning. I went round the head of the port — which runs for about 5 miles to the East^{d} from our anchorage terminating in 2 rivers, water nearly fresh at the head of the port. Placed marks preparatory to the Survey. Matalpies left this morning. 2 small canoes — here

[21] Robbers Nob.

only with 4 or 6 Indians — clams and Muscles appear their food; our people appear to be taking a lesson from them — for I observe them all collecting clams — and the Doctor tells me they are delicious — I am content to take his word for it — found the crayfish or lobster here but very small — scarcely larger than a good sized prawn.

MAY 2. Employed surveying the Port with Galley — all the other boats still absent — the day fine but the strong Westly & SW wind still blowing. The few natives here bring deer despite the game laws and their being out of season — one large one weighing 80 lbs bought today for 2 shirts value about 8 Shillings but the meat is by no means good. A few grouse are killed — but no fish have been procured here — which were so plentiful at the Entrance of Discovery passage — the variation of the compass observed today on shore — 22° 20' E.

MAY 3, THURSDAY. The weather very boisterous today a strong SWly wind which blows from NW or WNW thro the Strait outside and I fear is delaying the work of the detached parties; it was with difficulty I could get on with the survey of the port sheltered as it is — the tides run from 2 to 3 knots inside — as the full moon — advances.

MAY 4, FRIDAY. Got forenoon observations but the weather came over cloudy in the afternoon and prevented my getting corresponding ones — employed putting this port on paper. Got stars in the first watch. Mayne returned with gig — having brought the survey from Seymour Narrows to Knox Bay.

MAY 5, SATURDAY. A very fine day with NEly wind which brings fine weather everywhere on this coast. Got Equal altitudes and star latitudes — all the boats returned late this evening having encountered very strong winds and unpleasant weather during the last week and had some awkward times among the tide rips which get up without much warning causing a sea very dangerous for open boats and making great caution necessary — the two whalers stove badly were hauled on the beach to repair.

MAY 6, SUNDAY. An unpleasant day with fresh NWly wind and cloudy weather — some of the officers who accompanied by Indians have been on shooting expeditions during the past week report some very large lakes close to the head of the port ahead 2 days journey from them.[22] There are plenty of Indians and Salmon there so say our natives here. I regret that some inland journey has not been undertaken but all the surveying Officers have been so fully occupied that it has been impossible to carry out anything of the kind — moreover the weather is very unfavourable for camping out without a tent & gear which I could not spare parties to carry. A very remarkable mountain rises over the head of this port perhaps 20 miles off — it resembles a hat somewhat — and I have called it Hat Mountain.

MAY 7, MONDAY. Rain & SE wind. Mr Pender with gig went on in advance to examine the north side of Johnstone Strait. The 2 whalers repairing. My boat surveying outer part of port & Shark Sounding it — but obliged to return before noon in consequence of the rain which was constant during the day.

MAY 8. Rain and thick weather. Went across the opposite side of Strait and succeeded in Measuring a Micrometer base as a check on the triangulation — used a 20 feet beam extending the angle with an Az Alt instrument — Bull also outside in Whaler. Both boats returned at 4 PM. Sounding the Entrance of the port which is very narrow and Shallow. Heavy rain in the afternoon. The Natives left the port as soon as the rain commenced yesterday and returned I presume to their more permanent habitations — I fancy when rain sets in here surrounded as we are by such lofty mountains — it is likely to last for some days. Pender returned at 8 PM. having discovered a large port[23] 8 or 10 miles westward on North Shore of Johnstone Strait which will delay our arrival at Fort Rupert probably for another week.

MAY 9. Very gloomy. Mist hanging about the hills almost to the horizon — Baro low 29.70 — Whalers — under Pender and Mayne left to survey Westward — my boat finishing outside of harb^r. A rock with 4 feet on it discovered in the Centre of the Narrow entrance — it is marvellous how the ship escaped it in entering — the rock found after 3 boats

---

[22] Probably Fulmore Lake and Tom Browne Lake.

[23] Port Harvey.

have as we believed thoroughly sounded the passage. This together with the general shallowness of the entrance detracts considerably from the value of this otherwise fine port — there is notwithstanding a good passage in for Steamers drawing 18 feet at low water — tho it is narrow. The rise and fall of tide here at Springs is upward of 14 feet.

MAY 10. Got observations for rates this morning preparatory to leaving tomorrow — but rain and cloudy weather prevented my getting corresponding alt^{s} in the afternoon. The Season here is certainly a month later here than at the south end of the I^{d}. We have scarcely yet had a day of summer — with a Strong wind from NWest or SE — or continued rain, both unfavourable to our work. 2 or 3 canoes of natives appeared again today and my cloak was stolen from the Galley while ashore 200 yards from the ship by some adroit thief who saw us making preparations to be off. We have met with so few instances of theft indeed I can scarcely recall to mind one that our people are grown careless & boats are left without Keepers. One can scarcely be surprised at a robbery of this kind the poor wretches are miserably clad and my cloak made by Gillott — more than 15 years ago will no doubt be a comfort to some poor wretch tho the loss of it will be felt by me being an old boat Companion — Pender not back in the whaler as I expected he would have been. Barometer still low and not much inclined to rise — Snow falling. The tops of the low hills are covered within 1000 feet of their base since noon today altho none has fallen in that shape with us.

⚓ PORT HARVEY JOHNSTONE STRAIT

MAY 11. Left Port Neville early this morning and Sounded thro the Strait westerly a fresh West wind and thick gloomy weather. At 8 entered a port about 9 miles west^{d} of Neville & steamed in for 4 miles or less anchoring at the north head in a Snug cove in 9 fathoms. Whaler rejoined an hour after — I commenced the Survey of the port at once — which I named Port Harvey after Capt Harvey of Havannah, it is a very extensive place the north arm on which we are is about 3½ miles long. 1½ miles from its head a branch runs ENE for 3 or 4 miles and then turns NE for several more. 2^{d} Whaler Mr Mayne returned tonight in consequence of the boat having sprung a leak or rather half a dozen leaks.

MAY 11. The Matilpir tribe of Indians inhabit the Country around here: at war with the Hyders or Queen Charlotte Islanders who frequently come down to Victoria in their canoes and in passing generally have skirmishes with the different tribes as they pass. These Indians are if possible more filthy and dirty then their neighbours. They never appear to wash; and rub the fat from dogfish & deer into their skins as a protection from the Cold and to prevent Mosquitoes from Stinging; plenty of Fish and deer bartered for Shirts, tobacco etc. (Gowlland 1859–1864).

MAY 12, SATURDAY. All boats at work but 2d Whaler — repairing. From what I see of the extent of the port — today — it will take us 3 fine days hard work to complete it. Came on an Indian Village in the NE arm — regularly stockaded — very large Salmon drying arrangements round it, there don't appear to be very many natives here at present. They are a harmless miserable ill clad race — somewhat given to thieving which one cannot be surprised at. This has been a fine day tho cold. The season here is certainly a month behind the Southern end of the Island.

MAY 13, SUNDAY. May. A cold gloomy day with SE wind — walked from the north head of this cove — in a due west direction for 1½ miles and came upon the Head of a creek in Call Inlet — our course was thro a most remarkable ravine resembling the bed of a river which we at first took it to be until we found the water salt. Huge bolder stones were strewed over this gully or ravine which varies from 100 to 150 yards in width almost a dead level — so that at high tides the salt water has a fair passage thro, navigable for canoes — the pines growing on either side to the edge of the gully gave it the appearance of an avenue on looking thro it from this anchorage — which you may do for a mile or nearly so in a straight line. Returned at 3 PM — weather somewhat clearer. A splendid Snow mountain about 6000 feet rises immediately over the head of this north arm.

MAY 14, MONDAY. All boats away early. Mayne outside with a Gig to work westward to Port Bauza or Beaver Harb[r] [Beaver Cove] — Pender with whaler and Shark surveying the arm of this port which runs to NE. while my boat has been doing the outer part of the Harbour and ensuring the connexion between all the detached portions being accomplished under many different parties. The port is very extensive and so many narrow arms and thickly wooded Islands make it a difficult piece of work to carry thro expeditiously and plot. After leaving it the water is generally very deep — indeed our present anchorage is the only convenient one we have yet come across. Officers and men are in first rate working trim and the amount they get thro astonishes me — The natives bring Deer and Elk in any quantities and the people consume an amazing quantity and salt in still more. I don't think myself that the meat is very good — but anything goes down with them, they have good digestion and strong Stomachs — I don't think I have ever seen as much work done in the same time — as has been accomplished by our parties during the last month. And all good work which will stand any test; the Barometer has gone down this evening but our weather here is so much governed by local circumstances that the Baro. is scarcely any guide — the general thing is a Strong wind thro the Strait generally from Westward — today from East[d]. The high mountains which rise almost abruptly over both sides of the Strait deny us any clear weather — and ones temper is sorely tried when watching for observations of either sun or Stars.

MAY 15, TUESDAY. All boats at work. Day fine and very Sultry. Wind SEly. Sun extremely hot in the afternoon. Saw the first Mosquito for the Season — got equal altitudes and completed the East arm of the Port. The few natives here continue to bring deer and fish in quantities. One or two Elk have been bought and several Grouse. There have been no instances of theft here perhaps because we keep a better lookout. The fish are excellent. I think the true Cod. Certainly we have found better fish since we have been this side of Cape Mudge than to the Southward. These Cod are caught with hook and line by the natives. Our people have not been very successful; there is very little to interest the Naturalist; barnacles and Sea weed alone — nor has the dredge brought us any thing. Unable to get stars tonight — a clear night among these mountains is a rare occurrence.

MAY 16. Got observations for rates of the chronometers — boats returned in the evening having completed the Survey of the port which we believe now to be that marked on Vancouver's chart — as Call Canal — but the entrance of it is so erroneously placed that it is only by the general likeness so far as we can judge of it by his small scale that we arrive at this conclusion.

❧ NIMPKISH RIVER JOHNSTONE STRAIT

MAY 17. Gloomy morning — sent Shark to Nanaimo — to wait our arrival — doing a weeks soundings by the way. Started at 4.30 am — 2 Indians who came with us pointed out the position of a rock ½ a mile off the western entrance pt. of the Port. Ship struck soundings in 16 Fathoms — lowered a boat & sent one of the Indians in her but could not find less than 16 fms. However I have no doubt the rock is there. At this time of the Year there is no kelp and in such deep water it is very difficult to pick up any small rock such as this is described to be. Steamed to the westward. Picked up Gig Mr Mayne at 10 am just West^d^ of Beaver Cove which looks a snug little anchorage. At 11.30 passed Nimpkish river an extensive Stream which runs thro a valley leading to West Coast. The Natives say it is navigable for 2 days for canoes & then one days walk fetches Nootka — or one of the arms of it. This river flows from a very large lake[24] in the Centre of the I^d^ as well as 3 or 4 others — saw Vancouver's — Cheslakees Village at the Entrance of the River the bank is the same as represented on his sketch but the populous Village as there represented has ceased to exist — a few frames of houses only remaining. Cheslakees is gathered to his fathers — but his sons or descendants — I am told remain.[25] Passed his Vancouvers Sandy I^d^ which is composed of large Bolder Stones — and the water extends shoal

---

[24] Nimpkish Lake.

[25] On July 20, 1792, Captain George Vancouver met Chief Cheslakees at the village which he recorded as "Whulk" (Archibald Menzies noted it as "Wannoc") at the mouth of the Nimpkish River (Vancouver 1984: 625). When Richards arrived almost seventy years later, he found house frames. It is likely that the 'Namgis people were simply away at the time, having taken the house planks to their eulachon fishing site at the head of Knight Inlet. During the 1860s, the 'Namgis moved their winter village to Cormorant Island.

for some distance off it — all the way from the River to Rupert there is a moderate depth of water and Anchorage. We passed over two bridges of gravel — one with not over 4 fms at low water — and Anchored in Port Rupert at 2 PM — which is a pleasant place with a considerable quantity of level land about it. A pleasing Contrast to the Mountainous Region we have lately been sojourning in. Mr Waynton in charge of the Fort assisted by our old friend Cap Mitchell of the Recovery.[26]

§ FORT RUPERT–BEAVER HARBR

MAY 18. Sent the two whalers to connect the work with where Mr Mayne left off at Beaver Cove — the remaining 2 boats surveying the port.

I got observations for rates etc. on the beach at the west end of the Native Village. The landing here is bad everywhere. In the bay a long flat extends off at low water and the only place to land is on a beach or point covered with large Bolder stones — requiring great care to beach a boat. The port is exposed to Easterly winds and I am told that much swell sets on with a SEd — when however there would be comfortable anchorage westward of the Shell Islands. There is a very large Indian Village here, but few natives at present. They are away curing the small fish we call smelt or Hulicans.[27] Mr Waynton estimates the entire tribe to amount to less than 500 men when all present — the name of the tribe, Quoguellas or as sometimes pronounced Coquells.[28] He also informs me that there is a considerable amount of cleared land a mile from the head of Beaver Cove about 20 miles Eastward of this port ascending a rapid stream for a mile. He estimates the plains are 12 miles in extent. Deer Elk and Black Bear — Beavers very numerous. The Grizzly Bear not uncommon. I saw one Indian who had been very severely wounded by one of these

---

[26] Fort Rupert was established in 1849 by the Hudson's Bay Company on Beaver Harbour, south of Port Hardy. The Company was attracted by the presence of coal.

[27] The eulachon (*Thaleichthys pacificus*) is a small fish, sometimes called the "candle-fish" because it is said to be so oily that it will burn like a candle. Many Northwest Coast people who had access to eulachon runs rendered them for oil. Known as "grease," this oil was, and still is, a highly valued food item among Northwest Coast people. The fish can also be smoked.

[28] The Kwakiutl moved to Fort Rupert at Beaver Harbour in 1849 when the fort was first built. Their name for this place is *Tsaxis*.

beasts. This old man was between 60 and 70 years, very old for these people who rarely live beyond 50 — he looked 90. Heard of several harbours from Mr Waynton & Cap Mitchell; Bull harbr on Galiano I^{d}[29] Shucartie near this — & Sea Otter harbr on west coast near Cape Scott.[30] There is said to be a considerable quantity of land available for cultivation between this and Nimkish river — between it and the west coast; it is only a 5 hour journey from here to one of Quatsinough arms on west coast.

Succeeded in finding some of the Yellow Cypress[31] trees about ½ a mile from here & sent carpenters to cut them. They made extremely good boat plank the only wood on the Island adapted for it. I believe it is scarce on Vancouver I^{d} not growing except at its northern end. I remember it in Sulphur's voyage very plentiful at Sitka and other parts northward of this.[32]

MAY 19. Employed surveying the harbour — day gloomy. Rained all last night which I passed in my tent ashore with hope of getting star observations — failed in our yellow cypress trees — where they were first found — being too small & too great a distance from the ship — found some however nearer. The leaf differs from the white cedar — very much, being rounded, something like our cypress in England, and both sides alike — the white cedar or cypress more properly has a flat leaf the under part of it being white.

MAY 20, SUNDAY. A fine day but strong NW wind set in suddenly at noon — the white people with their Indian wives came onboard to church. Landed afterwards & looked at the cypress trees.

---

[29] Bull Harbour is on Hope Island.

[30] Gowlland repeatedly complains about the poor navigational information provided by the Hudson's Bay Company officers: ". . . they are the only people that are at all acquainted with the coast and anchorages and when asked for their knowledge it generally ends by them knowing nothing after all or their information so vague that we have to find out our own anchorages and anything else we may require to know of the Coast or Country" (Gowlland 1860–1863).

[31] Yellow cedar (*Chamaecyparis nootkatensis*).

[32] Richards visited Nootka Sound with Captain Edward Belcher on the H.M.S. *Sulphur* in October 1837.

MAY 21. Employed surveying the harbour. Got one cypress tree down about 20 inches in diameter — cut to 20 feet in length — the trail bad — heavy work but we must have them. Got star observations on Shell I^{d} at night.

MAY 22. Surveying the harb^{r} — observing for dip & variation on Shell I^{d} (Bull). Got off 3 lengths of cypress yellow — and if I get observations tomorrow shall be ready to leave on Thursday morning. Got star latitudes.

MAY 23. The Baro has been down to 29.70 for the last 12 hours and the weather has been foggy with northerly wind but clear overhead. Got all the observations I required today & moved our tents and tide party on board — sent Gig down the coast towards P^{t} Michael to get angles and meet us tomorrow morning. Entered an Indian lad here — son of the old Chief Wa-wa-te and took old Cap Mitchell of the old Recovery on board for a passage South — the night fine and clear — glass rising. The day has been broiling.

MAY 24. Left Beaver Harb^{r} at 4:30 am and steamed eastward thro Johnstone Straits. Picked up 2 of our boats off Pt McNeal. A long spit extends easterly from this point with no more than 20 to 24 feet on it at low water. A ship should get Haddington I^{d} onboard and pass along its shores about 2 cables lengths.[33] Port McNeal small but affords good anchorage. Passed Nimpkish river at 9 am. Vancouver's Sandy I^{d} should be given a berth — I can't discover why it is called Sandy I^{d} unless it be because it is composed of large bolder stones — an excellent bay on Cormorant I^{d} just abreast Nimpkish river anchorage in 7 fms. 4 feet on bar of Nimpkish at low water. 18 feet inside but only room for a small craft to lie — rise & fall here 15 feet at springs tides. Regular in Johnstone S^{t}. Flood stream runs East^{d} for nearly 2 hours after high water — and ebb — West^{d} for same time after low — strength abreast Nimpkish at Springs from 3 to 4 knots. Met 19 Hyda or Chimpcian canoes returning north — probably from Victoria. At 10 am whaler rejoined off Beaver cove — which does not afford any comfortable anchorage; wind SEly all day & fresh.

[33] Andesite granite from Haddington Island was used in the construction of both the Parliament buildings and the Empress Hotel in Victoria.

❦ BLENKINSOP HARB[R] JOHNSTONE STRAIT

Sounded down the Strait and anchored in Blenkinsop Bay 2 miles East[d] of Port Neville at 5 pm in 11 fms. A ship should not stand too far into the bay which shoals suddenly and dries for a considerable distance from the head — on west shore is best anchorage — bank'd fires for the night.

MAY 25. Left Blenkinsop harbour at 3.30 am — the small I[d] off West Entrance pt should be kept onboard within half a mile to avoid a reef extending off the small I[ds] west of Hardwicke I[d] until the reef is better known. We have fixed the position of the one that is uncovered but the natives say there is one which does not uncover — further off. We shall find it when the kelp grows in July.

❦ DISCOVERY CHANNEL–SEYMOUR NARROWS

Kept down the South Side of Discovery Channel sounding — the flood tide with us — between Helmken I[d] and south side of Channel got as little as 18 fathoms — here the tide is rather strong the passage being narrow. At 8 passed Knox Bay and at 9.30 am P[t] Chatham off which there is a rock but not extending more than ¼ of a mile — Sounded down the Channel and found but little tide altho only 4 days after new moon — until within 2 miles of Seymour Narrows where we got the last hour of the Ebb at 11.30 am and could not steam it. Steamed into the northernmost of 2 bays on the Eastern shore but could not find any anchorage within a cable of the head — so waited her underweigh and out of the tides until 1.30 pm — sent a boat to sound — found we could anchor in 10 fms 80 yards from head of south beach of the bay — the north beach very steep. At 1.30 pm steamed out and passed thro the narrows at slack water 2 pm. The low water by the shore occurred about 12[h]. 20[m]. Therefore the ebb tide ran thro the Narrows to the northward for nearly 2 hours after low water — which I imagine is the general thing here.

It is something more than 2 hours farther west[d] thro the passage. The high water at Fort Rupert today would be about 4.30 pm, here at the Narrows about 6.30 pm or 2 hours later. Passed Cape Mudge at 4.20 with a flow of about 2½ knots in our favour and steamed on to pass Cape Lazo at the distance of 3 miles — which we did at 7.30 pm. Thence down outside Hornby I[d] — the coast of which we passed at 2 miles and hauled in to clear Sisters rocks — then shaped a course for Ballinac I^{d} sounding in about 120 fms — 3 times an hour.

NANAIMO HARBOUR

MAY 26. At 4.15 am entered Nanaimo harb^{r}, a splendid clipper barque the Helen of Bergen lying at the entrance ready to start full of coals. Received our letters and coaled the ship at once. Got equal alts for rates — found Capt. Franklin established here as a magistrate, having his family with him. The first step towards civilization at Nanaimo.

MAY 27, SUNDAY.

MAY 28, MONDAY. Fine day. Swung ship for local attraction. Sent boat to Nanoose to renew our mountain mark. Busy all day answering letters.

MAY 29, TUESDAY. Got observations for rates of chronometers. Boat from Nanoose returned. Afternoon exercised firing great guns at a mark. Practice very bad. Not to be wondered at. Rode to the meadows — some good clear patches but of little use to any one — as the land is low and in sections to suit no one — a man may buy 50 acres of land and he will get perhaps ½ an acre of decent land, the rest rock and swamp. The first discoverer ought to have the choice if a bona fide settler — of the little good land there is — a rotten corrupt surveying department is the Curse of a new Colony.[34] There is much rotten here.

STRAIT OF GEORGIA–LESQUITA ISLAND

MAY 30, WEDNESDAY. Started at 5 am steamed up Georgia St sounding — landed on Sisters Rocks and placed a permanent beacon, then sounded North west end of Lesquita I^{d} and steamed Easterly between it and Texhada I^{d} anchoring in a good bay on the north side of the former in 12 fms. Commenced at once the survey — day very sultry & misty, very much like our old smokes of July.

MAY 31, THURSDAY. All boats away — Pender to survey the SW side of Lesquita — Mayne the west or SW side of Texhada — for about 12 miles

[34] Joseph Despard Pemberton was the Surveyor General of the Colony of Vancouver Island between 1859 and 1864.

from this anchorage — Bull to do the North Side of Lesquita — which is broken up into literally hundreds of small I^{ds}, and myself in Galley to carry the triangulation easterly to the Eastern ends of Texhada and Lesquita. I found a strong and unpleasant wind from SE today but got down to the Eastern end of both Islands. Texhada is reported to be a lime stone I^{d} — on the seacoast I found nothing but the old Trap[35] — Greenstone with occasional veins of quartz — the I^{d} is rugged and the hills rise very abruptly — making travelling extremely tiresome — certainly on this side (the South), there is no spot fit for cultivation, although a good deal of vegetation in the way of vetches occurs in the valleys and among the stones but no depth of soil — pines — but inferior in size — the Pencil cedar[36] is plentiful on the Sea coast but rather stunted. I got one tree 12 feet long by 10 or 11 inches in diameter — probably many more equally good or better are to be had. Camped tonight in a Snug cove on Lasquita I^{d}. Mosquitoes rather troublesome.

❧ LESQUITA–FIREWOOD BAY

JUNE 1, FRIDAY. Went this morning to Cape on P^{t} Upwood of Vancouver to SE pt. of Texhada I^{d} — got a glance of the Entrance to Jervis Inlet, and although doubtless there are few if any anchorages in that deep arm — there appears to me a probability of moderate depth — between Texhada I^{d} and the Main; I only judge from the formation of the latter — wherever there are clay cliffs I have generally found bottom in a reasonable depth. The weather has been very thick all yesterday with the stormy SEly wind and we saw none of the mountains or distant marks which I desired to see; today the wind was NWly which I think is the usual direction. It was also gloomy and hazy until 3 pm when it cleared and I saw very distantly Nanaimo Mount [Mt. Benson]. Mounts Moriarty and Arrowsmith thickly covered with snow on the 1st of June — returned to the ship at 6 pm, convinced that neither on Texhada or Lasquita I^{ds} is there any land suitable for the settler — or indeed

[35] Trap Rock is a dark-coloured igneous rock more or less columnar in structure.

[36] The "pencil cedar" is a juniper. It was formerly considered the same species as Rocky Mountain juniper, *Juniperus scopulorum*, but is now designated as a separate coastal species, *Juniperus maritima*. Juniper wood is used to make pencils as well as the famous "cedar chests" (Adams 2007; Turner 2010).

likely to be profitable to any one. The poverty of the soil is indeed a most serious drawback to colonization in this dependency of Great Britain. I have never seen in any country in the world such a general absence of moderately good land as I have met with both on the Vancouver I^{d} shore and that of British Columbia. We have seen no natives here; indeed I don't see what is to bring them here.

JUNE 2, SATURDAY. Got true bearing and measured a base with a 20 foot board. Measuring the angle with a small Az alt inst.[37] I find I can get a very correct base this way and measure a mile that is to say sufficiently correct for a harb^{r} or a small piece of work. Employed the afternoon in laying down the work. Mayne returned this evening having completed his work for 12 miles above Texhada I^{d} shore. Shark also returned.

JUNE 3, SUNDAY. A fine day. Summer has only commenced since the 1st of June.

JUNE 4, MONDAY. Boats away — Mayne carrying the survey round at NW end of Texhada, Gig Mr Browning round SE end. Shark sounding round this I^{d} of Lesquita — I went on the Summit of the Nob a remarkable excrescence very much like Trematon Castle in the Tamar — and about 1000 feet high rising in the middle of the I^{d}.[38]

Started at 7 a.m. & reached summit at 9.30 after a tough walk over very hard & rugged ground from fallen trees, stones and undergrowth. Passed over several stoney ridges of about 500 feet elevation & then down again into valleys nearly as deep. Some very fine cypress timber — what we call the white cedar — saw no living animal. I have rarely seen any part of the coast so extremely destitute as this. Day exceedingly warm. Mosquitoes and butterflies on the summit. Remained there 4 hours getting angles. The atmosphere has been any thing but clear these last few days as it rarely is in June & July. Perhaps from some distant smokes of the natives. Returned on board at 4 pm. Pender returned in 1st Whaler at 8 pm having completed his work on the outside of the I^{d} and found anchorage in 2 places at the NW end. The Island is exceedingly broken &

[37] Azimuth and altitude instrument.

[38] Trematon Mountain. Named by Richards after Trematon Castle, a late-eleventh-century castle in Cornwall, where Richards was raised.

cut up with numerous islets on both sides. It is 10 or 11 miles long by 3 to 4 wide — no anchorage on the inside except the one we are at — which is good enough.

JUNE 5, TUESDAY. Laying work down all day. 3 boats absent. 3 canoes visited us today and brought some fish. They set fire to a small island which produced so much smoke as to obscure the whole country round for several hours — and I can easily believe that the general haziness experienced in summer is attributable to these smokes. The natives say they do it to catch the Deer.[39] Cut several of the pencil cedar trees on Texhada today. They are small but the wood is very good. I got one 12 feet long by about 10 inches in diameter. Mr Brockton shot 2 bald headed eagles today & brought me one exceedingly fine specimen which I shall have skinned. Tomorrow morning I propose leaving this anchorage and seeking another on the North Side of Texhada. We have seen nothing like limestone on these I^{ds} altho it was reported that Texhada was entirely of that formation.

MALASPINA STRAIT–NANAIMO

JUNE 6, WEDNESDAY. A strong NW wind sprung up during the night which made the anchorage uncomfortable tho not unsafe. Weighed at 4 am and steamed East between the two Islands. Hauled round P^{t} Upwood when the wind blew so strong thro Malaspina Strait that we could not steam it — water very deep nearly 200 fms. At 9 bore up stopped engines and ran under sail for Nanaimo, there being no nearer anchorage to leeward and not being able to steam to windward. At 11 anchored at Nanaimo and found a mail. Strong NW gale all day.

JUNE 7, THURSDAY. Gale blew all night and continued so fresh all today that we were unable to move. HBCo Steamer arrived from Victoria on her way north from whence she had returned prematurely in consequence of the illness of Cap. Dodd — who died shortly after landing at

[39] Burning was a common resource management practice on the Northwest Coast. For indigenous people everywhere, burning was used to maintain grasslands and clearings, and to reduce brush in woodlands (Turner 2005: 152). In addition to providing habitat for deer, berry yields were enhanced by such management practices as burning, pruning, and possible reseeding on cleared plots owned by individuals and households (Deur 2002: 16).

Victoria. Cut a yew tree for keel for 2nd Whaler — and did a little work in Departure Bay. Several canoes of Chimpseyans arrived.

❧ TRIBUNE BAY HORNBY ID–NANAIMO

JUNE 8, FRIDAY. Moderate but gloomy morning. Started at 4 am and steamed for Texhada — passed up Malaspina St, 46 canoes passed Southd on the western side of the Gulf this morning. Water very deep in Strait nearly 200 fms. At noon fired guns to attract boats attention. 1.30 Gig & Whaler returned the latter badly stove & lost rudder in late gale — then provisions up last night. Passed round west end of Texhada not finding a harbr said to exist by HBCo. authorities named by them "Beware". There is an indentation on the Id shore but had shelter & amp water. Passed between Harwood Rock always uncovered and a covering rock which lies about ½ a mile off the W pt of Texhada. Least water 20 fms. Boats just returned report the west side of Texhada is composed of lime stone & brought good specimens. About 3 miles from the West end lime stone commences. This end of the Id is low and more favourable in appearance than the other parts which are barren & rocky to a degree. Picked up the Shark at 3 p.m. and steered across for Hornby Id. anchoring at 7.30 pm in Tribune Bay. This anchorage affords a convenient depth 6 to 9 fms, but is open to SEly winds which however I don't think would ever send in sufficient sea to endanger a vessel with good ground tackle. A sunken rock lies on the West side of the Entrance and the Eastern shore should be kept onboard in entering or leaving.

JUNE 9, SATURDAY. Left Tribune Bay at 4 a.m. Stopped off at St. John Islets which lie off the East [low] pt of Hornby Id & landed to get angles. Proceeded on to Ballynach [Ballenas] Ids and landed there at 9.30 am for same purpose. Atmosphere hazy with fresh NW wind. Did not render the observations very satisfactory. There is rarely a clear atmosphere between May and Sept. I mean very clear, unless after rain or with a SW wind. Anchored at Nanaimo at noon and to my surprise found I could get no coals, as I was promised I should before I left 2 days ago. Labuchere HBCo steamer using the 2 launches. We only wanted 17 tons which we could have taken in 1½ hours, and as I did not care to wait until Monday with the uncertainty of getting them then, I at once weighed and proceeded outward, intending to visit San Juan. It does not seem very unreasonable to expect that a ship which is constantly employed in surveying for the benefit of all should be supplied with a few tons of coals promptly — by a large Co. which are considerably benefitted by her labours. Indeed most merchant ships would under such circumstances forgo an hour or two or at any rate share the opportunity of getting the coals onb^{d}.

COWLITZ BAY–SAN JUAN–ESQUIMALT

JUNE 9. At 11 pm we anchored in Cowlitz bay, Waldron I^{d} in 8 fms. This is a good anchorage and easy to pick up at night if not very dark.

JUNE 10. Left at 4.30 am and proceeded thro Spring passage into Middle Channel and at 7.50 anchored at San Juan. Sunday morning, landed, saw Mr Griffin HBCo agent and was called on by Cap Pickett, USA comdg the company of US troops which occupied the S^{o} end of the I^{d}. Cap. P. very polite & pressing for us to remain.

JUNE 11. Left at 5.30, passed out of South Entrance, boats sounding on the Salmon bank. Ship also looking for a shoal on which it was reported Satellite struck some time since. At 3 pm having thoroughly sounded the bank and found no shoal that was not already shown on the chart, proceeded for Esquimalt & anchored at 6 pm. Found the admiral was absent on Satellite visiting the Marine Camp at San Juan I^{d} (Roche harb^{r}). He returned at 7.30 pm. 10 pm Shark returned from Nanaimo.

JUNE 12. Commenced the repairs of our boats which were much damaged by the late survey in Discovery Channel and Johnstone Strait.

JUNE 15. The U.S. Store Ship Massachusets came in. Her Comdr reported having struck on a rock on Salmon bank off San Juan where we had just been sounding. I dispatched the Shark to look for it on the 18th; she returned on the 20th reporting that the rock was a bank of stones on the bank already shown on our chart with 2 fathoms on it. The ship had no business there.[40]

[40] According to Mayne the *Shark* found the rock "exactly between a line of soundings of Penders and one of mine, shewing again how impossible it is to make certain, however carefully a place may have been sounded" (Mayne 1857–1860: June 18, 1860).

❧ FRASER RIVER–NWESTMINSTER–DOUGLAS–HOPE

On 16 June I went with the Admiral, Cap Fulford & steamed to the Fraser in Stern Wheel Steamer Govr Douglas. Visited NWestminster, Douglas & Hope and returned on 21st. Not much impressed with the idea that the country was making any rapid progress, or that the difficulties of navigation of the river were chimerical as I have heard asserted by the B.C. colonists. The steamer had 11 knot power and in many places scarcely made any progress. River intricate and dangerous. Douglas is at the head of Harrison Lake 42 miles above its junction with the Fraser. It is a Miserable Village composed of grog shops & Billiard rooms, the inhabitants smart Yankees principally. The place only important as the starting point of the trail to the upper Fraser by Lillooet & Anderson Lakes where a road is now being made with great labour & expense by the R.E. [Royal Engineers] under Cap Grant also by private contracts with Civil Surveyors. From the confluence of the Harrison to Hope is 32 miles which took us 12 miles [hours] to accomplish in the steamer. Hope is a very pretty spot built on a plateau or flat under mountains of 5000 feet. Scenery extremely grand. From here a trail is being cut to Similkimene Valley[41] said to be rich in metals.[42]

❧ HARO STRAIT–BURRARD INLET

JULY 31, TUESDAY. Left Esquimalt at 4.30 am — Termagant and 'Alert' in company for Nanaimo.[43] Passed inside Zero rk and up Sidney Chan-

---

[41] By 1860, the focus of the gold rush had shifted from the Fraser River to the Similkameen River and to the Cariboo, north of Quesnel, where it was said that a person could make up to $400 a day (Akrigg 1977: 191–193).

[42] Richards gave instructions to Mayne to conduct an overland journey from the head of Jervis Inlet to the interior, in an attempt to reach Bridge River. The "first and greatest object of [the] mission is to discover a practicable route by which supplies may be conveyed into the country, the want of which and the enormous rate of Freight, are now the great drawback of the prosperity of the Colony" (Richards 1860: June 30). Mayne's eigthteen-day journey took him to Jervis Inlet, Pemberton, Douglas and Hope (Mayne 1857–1860: July 21, 1860). (See Mayne 1862: 190–206 for a full account of this journey).

[43] The ships were dispatched to Nanaimo because settlers in the region were concerned about the number of First Nations people who were arriving there. Newspaper reports suggested that a thousand people had arrived in recent days (*British Colonist* July 31, 1860, August 1, 1860 and August 3, 1860).

nel — between Canoe rks and Turnbull reefs, steering for Active Pass. Ships formed in line ahead. At 11.50 the ships closed up and at noon entered the Pass, Plumper leading. The flood stream with us, about 4 knots, perhaps less. I intended to have entered the pass at slack water but Alert's bad steaming delayed us until ¾ of an hour after the stream had commenced; the ships all turned into the pass with great ease — which in its western entrance is about 600 yards wide, on making the turn out of the pass I observed that Termagant did not come to — to port and saw that unless she answer^{d} her helm shortly she must run on the pt outside Laura [Laura Point]. She just escaped running stern on to the rocks & took them about 10 feet under water on her starbd bilge under the fore chains, carrying away the trees with her fore yard and shoved off without stopping. The shock is described by those onbd as being very easy and no damage was believed to have been sustained beyond rubbing off copper. About 20 minutes afterward she signalled something wrong with rudder apparatus. I went along side & found the helm would not go more than one turn a starbd. She is fitted with Adm^{l}. Martins patent lever steering apparatus; by knocking out the wood lock the rudder went over hard a starbd but not I fancy so freely as it should. It is difficult to account for this mishap. Cap Hall says the helm was hard over starbd. But that it had no effect on the ship — she sits 4½ feet by the stern drawing 19 feet aft 14½ for^{d}. An undercurrent may have existed which had effect on her long keel and not on her fore part. She did not enter the pass at full speed. When it seemed inevitable that she must go on to the point Cap. H. stopped her and in that moment she ans^{d} her helm — he turned ahead full speed & she avoided giving the rocks her stern — the pt is steep to an almost up & down sandstone cliff. I have frequently been thro this pass at all times of tide. 'Satellite' and Pylades both as large & longer than 'Termagant' have been thro without difficulty; they sit on or almost on an even keel — & I would not hesitate to take the largest Steam line of battle ship thro; where Termagant struck the width of pass is 700 yds. Cap Hall describes Termagant as steering generally very well — she used to have the reputation of steering badly I am told; after this accident we steered for Burrard Inlet anchoring in English bay in 9 fms at 6.30 pm. The Active Pass saves 11 miles and makes it easy to reach an anchorage by dayl^{t} from Esquimalt and thus avoid the risk consequent on knocking about at night among strong tides.

AUGUST 1. Wednesday morning I started to look for my detached parties leaving Mr Bull to bring 'Termagant' and 'Alert' to Nanaimo on the morrow; before leaving I went onbd Termagant & Cap H. informed me he suspected the ship was making water. Proceeded on Plumper past Passage Id and into Howe Sound by Bowen passage and at north eastern entrance and on along shore towards Jervis inlet; at 2.30 picked up our boats and at 4 h the Shark; steered for Nanaimo against a strong SE wind. Kept over towards Ballinac Id and steamed up on the western side of strait where the force of the wind is less — at 8.30 we moored in Nanaimo.

NANAIMO–ESQUIMALT–NANAIMO

AUGUST 2, THURSDAY. Found a Wesleyan Chapel rising on the Town Hill — and other improvements in the erection of new houses and making of trails — a large number of Haiders here, among them the Chief Edensawe with his family. Bought 6 silver bracelets from him which he unceremoniously took off his wife's arm. 4 dollars each. They are made out of a dollar with the crest of the tribe impressed on them. Placed 2 boats, one on the point (Gallows) & one on South sand, for ships to enter. At 2 p.m. they arrived and moored in the harbour. Found Capt. Hall's suspicion confirmed the "Termagant" was making 3 inches of water an hour; Got our coals out.

AUGUST 3, FRIDAY. I decided to go to Esquimalt and see the Adml about "Termagant" & to bring the Diver up with me; to have sent a boat would have involved a great loss of time so I started at once, and steamed down the Main Channel anchoring outside Esquimalt at 11.30 pm. I did not go in as nothing could have been done that night.

AUGUST 4, SATURDAY. At 7 am steamed in and communicated with the admiral. Recd. Ganges Diver & gear — also Mr Ford the Carpenter. At 1 pm left again, picked up "Forward" gun boat off Victoria & took her on; by admls order — steamed all night & should have reached Nanaimo in mid watch but "Forward's" tubes got cracked & had to stop 4 hours.

NANAIMO–TRIBUNE BAY, HORNBY ID

AUGUST 5, SUNDAY. Moored in Nanaimo. Forward in co [company], at 7 am. Proceeded at once to examine Termagant's bottom & by noon ascertained that the outer planking had been gouged out for 4½ inches deep, 6 feet long & 6 inches wide by the point of a rock underwater. This had occurred between two planks thereby damaging the oakum forcing it out and causing the leak. Other injuries of a slighter nature, partial rubbing of outer planking and copper off, also sole piece of rudder, [illegible]; and it was determined to stuff the greatest injury with oakum & tallow, nailing fearnought & copper outside.

AUGUST 6–7, MONDAY & TUESDAY. Did this with the diver's help and the leak was diminished more than one half which was easily kept under with one pump, twice a day for ¼ of an hour each time.[44] Strong NW winds and the slow way "Alert" could get her coal detained me here till the morning of the 8th, when we left with "Alert" in co. steaming against a strong NW wind. Top masts on deck making very slow progress. The Shark is a great drag upon us and "Alert" has the advantages today. Wind continued fresh until near noon, when being West of Ballinac I^{d} and near the shore it fell light — weather very fine. A little snow still on Mt. Arrowsmith. Saw several large sharks which are said to make excellent oil.[45] All along the coast from Nanaimo west^{d} there is a large extent of low land wooded with [illegible] prairie a short distance within. It extends about 10 or 12 miles from the sea coast to the base of the mountain range; at 3.30 pm we anchored in Tribune Bay in 7 fms — "Alert" anchored alongside us. This is very good anchorage with all but SEly winds to which it is in some measure open tho I apprehend a ship with good ground tackle would under any circumstances lay in safety. There is perhaps 2000 acres of land on this I^{d}, the greater part available as grazing or agricultural land. It would be a very good investment in my opinion. The wood is neither large nor thick and the clearing of it would be attended with no great difficulty. It is probable that coal will

[44] The *Termagant* was sent to San Francisco for repairs but while being lifted into dry dock at Mare Island, she slipped and went onto her side causing great and expensive damage to the dry dock but none to the ship (*British Colonist* 1859–1862: October 9, 1860).

[45] The species described by Richards may be the basking shark (*Cetorhinus maximus*), a very large, plankton-feeding shark common in the Strait of Georgia until over-hunting and a federally sanctioned slaughter caused its disappearance in the mid-twentieth century.

be found here. Deer are numerous — the formation is sandstone and shale at the low water line. It would be a capital cattle station & good rich grass; also vetches or wild pea grow luxuriously. Much of the land is already clear.

❦ DISCOVERY CHANNEL–KNOX BAY

AUGUST 9. Weighed at 4.30 am and steamed round the eastern side of Hornby I^{d} for Cape Lazo carrying a line of soundings. "Alert" in company, at 8.30 passed Cape Lazo. SE^{d} of which extends a long Boulder reef for 2 miles — drying off for a mile at least at springs.[46] Steered for Cape Mudge and entered Discovery passage at noon the ebb being with us. It appeared but little after high water by the shore. This is about the neaps and it ought to have been high water at Cape Mudge at 8 am. However during neaps there is not so great a regularity of tide as at springs. Shortly after 1 pm we passed the Seymour Narrows with a slight ebb or favourable stream and proceeded on for Knox Bay distant about 26 miles; there is a considerable difference in the appearance of the country since our visit 2 months since — the snow has gone from all the lower hills tho it rests in considerable quantities about the summits of the second or more inland mountains from 6 to 7000 feet in elevation. The natives are now occupying the sea coast villages I suppose for taking salmon but none of them attempted to come alongside as they used to do; all our old marks standing — equally creditable to the natives & our own people; the latter for putting them up so firmly, the former for allowing them to stand; at 3^{h} we rounded Pt. Chatham, and at 5 pm anch^{d} in Knox Bay — "Plumper" in 13 fms 50 yds from low water mark. "Alert" in 23 fms ½ a cable outside us. Sent seining parties for fish and the men as is their custom lined the beach with fires.

❦ KNOX BAY–PORT HARVEY

AUGUST 10, FRIDAY. Weighed at 5.30 am after getting a few soundings in the bay and steamed to the westward against the flood sounding, the strength of the tide is not great in the wide channel but when passing

[46] This is likely Comox Bar and White Spit located to the southwest of Cape Lazo.

between Helmken I^{d} and the shore of Vancouver I^{d} about 7 miles from Knox Bay, we could only go ahead [a]gainst it (for a short time). The greatest strength of the stream occurs about a mile eastward of Helmken I^{d}; at 10^{h}. 30^{m} passed Port Neville, and at noon were off Port Harvey. Here I dropped 2 boats to look for a reef which the natives reported on our last visit, but which could not then be found; on this occasion however the kelp having grown it was easily picked up. Steamed into Port Harvey and anchored in our old berth. "Alert" came to, close outside us. This is an excellent anchorage in 7 fms; a few natives followed us in, hauled the seine but were not very successful, caught a few flat fish — afterwards about 100 salmon. Boats returned, found the reef with 5 feet at low water. A passage inside it.

AUGUST 11, SATURDAY. Weighed at 7.30 am found a strong westerly wind outside and a thick fog rolling in from the westward. Returned to the anchorage. It shortly cleared but blew fresh all day. Got equal altitudes — and T. Blaney observed a very large number of spots on the sun the appearance as seen thro an az and alt circle with refracting inverting Telescope. Great numbers of canoes came from Call Creek. Very dirty creatures among them. I recognized my canoe stolen from Esquimalt more than 12 months since. Blew strong from the NW^{d} all day.

AUGUST 12, SUNDAY. A quiet gloomy day. Landed for an hour but was glad to return on account of the Mosquitoes.

FORT RUPERT

AUGUST 13, MONDAY. Weighed at 4^{h}. 30^{m} am, "Alert" in co. Rainy morning and afterwards thick fog until 10^{h} am; picked our way along to the west^{d} sounding, "Alert" following; at 10.30 anchored in the bay on S^{o} side of Cormorant I^{d}, the flood running some 3 or 4 knots against us; this is a good anchorage, out of the tide, in 7 fathoms; the west end of the bay has a remarkable bare sandstone cliff by which it will be immediately recognized. Here I left the Pinnace and 2^{d} Whaler under Mr Mayne & Mr Blunden to complete the north sides of the group of Islands and sound the passages between them, provisioned for 14 and 8 days. At 11^{h}. 30^{m} am weighed and steered for Port Rupert distant 20 miles. I intended to have passed on the north of the small I^{d} Called

"Haddington", between it and "Malcolm" I^{d} but a very extensive kelp patch appearing in midchannel I passed South^{d} of the small I^{d} of Haddington. Off the SW end of Malcolm I^{d} we observed a kelp patch extending for more than a mile and giving it a berth — which I thought would clear all off lying danger we suddenly came into shoal water & saw the kelp streaming under water, on this patch there is 4½ fms at low water. We kept Vancouver I^{d} on hand after this and steamed over almost our old track up — being conscious that the wide expanse of water to the eastward was unsafe to traverse until thoroughly sounded. At 2^{h}. 30^{m} we anchored at Fort Rupert, "Alert" in co. On landing I found all the natives sitting on the beach outside the HB Fort gate about 600 in number. I immediately perceived that there was something wrong having before suspected as much from the chiefs not coming off to the ship; none of them got up to meet me on landing — as was their custom to do of old. I found from Mr Waynton that they were much alarmed by the visit of the two ships, particularly by their signalling to each other and anchoring in berth abreast each end of their village. Their consciences I imagine furnished these fears.[47] Mr Waynton informed me that a drunken debauch had been going on since their return from Victoria some two or three weeks since, that almost all the natives had been during that time in a shocking state of intoxication from the effects of which they were only now recovering, and which their appearance bore testimony to; at times Mr W. had felt very uneasy lest some drunken attack should be made on the fort — as the greater part of his own men had participated in the debauch. It appeared also that they the Indians had murdered 2 of the Southern tribe of Songhees and brought their heads up which were now hung up in triumph. As I had been applied to at Nanaimo to procure the liberation of a woman named Hoo-saw-I who

[47] Fear of the Royal Navy's "gunboat diplomacy" was well-founded. In 1850, the sloop of war *Daphne* was sent to northern Vancouver Island to apprehend the Newitti (Tlatlasikwala) who were charged with the murder of three runaway Hudson's Bay employees. The crew went ashore and amidst musket volley and gunfire from the *Daphne*'s pinnace, burned the houses and destroyed the village of Pakluntz (Gough 1984: 44; Galois 1994: 300). The destruction of villages was not infrequent on the coast, occurring numerous times at different locations during this period. In Victoria it was not uncommon for naval brigades to march through town in a display of military force intended to intimidate the First Nations who visited the capital (Gough 1984: 70).

these people had captured about a fortnight since, I thought this a good opportunity to assemble the Chiefs and remonstrate with them on the impropriety of committing murders and capturing people of other tribes. I therefore requested M[r] W. to tell the chiefs I would meet them tomorrow — they felt somewhat relieved but not altogether at ease on account of the presence of a second ship. I desired that nothing might be said about my intention to demand the woman but unfortunately M[r] Cooper Harb[r] Master who had been embarked in "Alert" to give any assistance to Cap Pearse considered that this matter came within his province and had opened the subject to the chiefs unknown to me, and which was productive of much embarrassment to me and great inconvenience to the service causing the delay of the "Alert" here some days. A rainy unpleasant night, wind NW.

> AUGUST 14, TUESDAY. This is the next Settlement of the Hudson Bay Company north from Nanaimo a very pretty little fort in charge of Waynton a young fellow who is most kind to all who go on shore. The Indian Village is much larger and in better condition than any I have yet seen — Indians cleaner — houses larger and Substantial with little Punch & Judy sort of Boxes inside fitted up for Sleeping Apartments. They number as 600 good fighting men all well armed and generally capital shots with the Rifle. There is a Capital Garden attached to the fort which produces almost Every thing. The soil is very rich — Currents; Gooseberries; Strawberries etc. in larger quantities than I ever saw in the best gardens in England (Gowlland 1860–1863).

AUGUST 14, TUESDAY. At 10 this morning I landed, accompanied by Cap Pearse and some of the officers. M[r] Cooper was also present. I had arranged previous to landing that the two ships should fire shot & grape at a target — I had been given to understand that they were very obstinate on this question and I thought it might hasten their decision to comply with my demand. I requested them to come inside the Fort enclosure, which they were evidently unwilling to do, and at least seeing that only a few would come I met them outside. I first told them that the governor was very angry at their conduct and explained to them that instead of taking revenge on innocent members of other tribes for injuries not inflicted by them, they ought to go to the Governor and make their complaint and that if the guilty parties could be captured they would be punished; that if they took the law into their own hands they — themselves would be punished. They had the old story and one difficult to combat — that it was Indian Custom that their people were murdered by the Southern tribes and they got no redress; that now an Indian at Victoria named La-Hart had killed their chief T-Coos-ma for which they had killed the two Songhies. The brother of T-Coos-ma was present, and with most earnest and impassioned manner declared that he intended to organize a war party & go down to take revenge on all he met. I told him that if the murderer of his brother could be identified he would be brought to punishment at Victoria by the Governor — he said if the Governor would hang La-hart he would be satisfied and would give me a large sea otter skin — but I plainly saw the spirit of revenge was burning within this man and I endeavoured to appease him and the rest by assuring them that the Governor was as much their friend as of any other tribe and that he was much grieved at T-Coos-mas death. They all feel this murder very much and say they will not be satisfied unless the murderer is brought up here and hanged. I strongly advised them to abstain from intoxicating drinks which were the ruin of body and soul, and also pointed out to them the necessity of adopting our habits as regards murderers. I fear the latter will be almost a hopeless piece of advice unless this interview is followed up by others — and they see we are prepared to adopt strong measures if necessary, which I assured them we were. One old Chief, "Whale", said, "What I said was good, but why did we not send a priest among them to teach them as we had done at Fort Simpson, alluding to Mr Duncan[48] who I believe is known throughout the land, fifty such men as him would have more effect than as many men of war. I reserved the hardest demand I had until the last and then told them I must have the captive woman, Hoo-saw-I; this occasioned

[48] Missionary William Duncan.

great commotion among them, and much remonstrance. They said would I get back all their slaves taken by the Nanaimo and other tribes. I replied I would try if they would give me names and particulars when I was immediately overwhelmed with such a host of cases as had occurred within their memory. I noted one or two in my pocket book and promised to do my best; for the woman they said she was 3 days journey off and had been sold for 50 blankets[49] which I ought to pay, and which I declined, and replied that if I did not get the woman I should be obliged to use force. All this time shell grape & canister were blazing away from both ships and I could plainly see had the effect of terrifying them — but they would not decide to give up the woman; at least after half an hour's delay, they went and fetched the man she belonged to who came forward in an apparently very ferocious mood. He had a pole or stick in one hand and a knife in the other. He then commenced an harangue of the most impassioned kind, stamped his feet, flourished his knife and told of all the wrongs his tribe had suffered from other tribes — I told the interpreter to tell him I admitted the wickedness of the other tribes and I was equally anxious to restore their slaves as this one if I could do so. I fancied once or twice he intended to put his knife into me, so ferocious was the expression of his countenance — but I believe he had no such intention; the wind up was that he and those about him declared that a hundred of them would rather die than give her up to the Nanaimos but that they would give her up to me; and then a second Indian who appeared to be the friend and second of the first took his pole or wand and after making an impressive harangue placed it in my hands declaring they had given me 100 blankets. I was told that the placing the stick in my hand was a solemn pledge that I should have the slave. They then asked for 3 days to fetch her assuring me that she was a long way off. They presently prepared a large canoe with six men but before starting to fetch her, asked me to pay the rowers for their trouble. This I did not object to do — but told them I should keep both ships here until they fulfilled their pledge. I believe but for Mr Cooper's interference I should have got her on the spot — but in consequence of it they

[49] Hudson's Bay blankets were an important part of the economy on the Northwest Coast. Most trade items were paid for in blankets or for a value measured in blankets.

were prepared and had either sent her away the moment he communicated the thing to them, or else pretended that she had gone — thereby causing me much embarrassment and annoyance. The fact is M^{r} C was desirous of making capital of this affair and from his little knowledge of the people he signally failed and involved me or was liable to have done so in serious consequences.[50]

> AUGUST 14. Employed during the forenoon firing Shot and Shell at a Mark on shore. Captain went on shore and had a long "wah wah" with the Chiefs and principal Men of the Tribe by aid of an Interpreter respecting a Slave they have in their possession and belonging to the Nanaimo tribe — the latter having made a petition to the Governor requesting his interference, and to restore them their tribeswoman. After a great deal of talk on both sides they agreed to give her up on condition of their houses being spared, which the Captain had threatened to destroy if they would not comply with his demand. A canoe was manned and sent over to some Island where they had concealed her on our arrival; and we ultimately got the poor wretch onboard; She appeared as much terrified at being rescued as when taken — thinking we were going to hang her or Something worse (Gowlland 1859–1864).

❧ SHUCHARTIE HARB^{R}–GOLETAS CHANNEL

AUGUST 22, WEDNESDAY. We remained at Fort Rupert until this morning. The "Alert" sailed for Queen Charlotte I^{d} on Friday morning the 17th. The slave woman Hoo-saw-eye was brought over on the 18th and received aboard the ship — having paid 6 blankets for the Indians to bring her across. During our stay at Fort Rupert the surveying parties were away — Mr Mayne with 2nd Whaler and Shark left at Cormorant I^{d} on way here — Mr Bull and Mr Pender carrying the triangulation

[50] In his reporting letter to Admiral Baynes, Richards did not mention Cooper's interfering role (Richards 1860: August 17).

westward; myself getting the metric angles in this neighbourhood and M[r] Browning completing the plan of Beaver harbour. Fogs impeded us considerably almost every day during our stay here. They have prevailed from early morning until noon to such an extent as to obscure all objects — and cause great delay. The prevailing winds are NWly and with these winds the fogs roll in from seaward where they are formed and not finding here the essential conditions to their existence they disappear as soon as the sun's power is sufficiently strong to produce such an effect. The "Otter" steamer anchored at Rupert on the 18th and left early on Sunday morning 19th for Fort Simpson. She had on bd M[r] Duncan — M[r] Tugwell and wife going to Simpson to take M[r] Duncan's place — and Miss Meesun a celebrated Vancouver I[d] traveler. At 6[h]. 30 am today 22[d] left Fort Rupert and steamed out by western passage — dropped Shark off Port Hardy, an open bay, and proceeded on sounding thro Nawitti on Goletas Channel — water very deep, no bottom with 180 fms. At 11.30 we anchored in Suchartie Bay — a very poor place water deep in outer part & shoaling very suddenly — anchored in 10 fms. Found we should tail on to a 10 foot bank — weighed again and anchored half a cable farther out in 17 fms — not recommended as an anchorage for sailing ships, well enough in summer time for a steamer. Bull went to examine Valdes[51] I[d] for a better harb[r]. M[r] Browning to make a plan of bay. I went along North Shore of Vancouver I[d] to mark it & prepare for the triangulation. Weather thick and gloomy. After going 3 miles found a heavy swell which prevented my landing. I suspect we shall have some rough work to get the survey round the North end of the I[d] in a proper way. Fogs would seem to prevail during August — north of this. I don't think, where we are, that they do to any extent in summer. Future observations must determine whether they do in October. I suspect they do.

---

[51] The Islas de Galiano y Valdes were named after Dionisio Alcalá Galiano and Cayetano Valdés y Flores, commanders of the *Mexicana* and *Sutil* who explored and surveyed these water in 1792 (Hayes 1999: 81, 93). They refer to a group of islands on the north shores of Goletas Channel. They appear on later maps as Valdes and Galiano Islands (Hayes 1999: 13) suggesting that Hope Island was known as Valdes Island. Initially Richards, perhaps mistakenly, refers to Hope Island as Galiano Island but he later differentiates Hope & Galiano Islands. In 1900 Galiano Island was renamed Nigei Island (Walbran 1971: 356).

AUGUST 23, THURSDAY. Galley and 2^{d} Whaler surveying the harb^{r} and neighbourhood. Weather very foggy nearly all day. Went across to Galiano I^{d} [Nigei Island] of the chart and got some stations. At 4 pm it cleared but too late for observations. Low water today about 11 am (neaps) measured a bar at head of harb^{r} of 1500 feet — and could not find any of the yellow cypress here as I had expected (native name te-ough).[52] Wood very dense, could not penetrate in land for many yards.

AUGUST 26, SUNDAY. All boats returned last night. The fogs in the morning and the surf that breaks as it would appear constantly on the western & NW sides of the Island as well as on the Coast of Vancouver I^{d} as soon as we get 3 miles west^{d} of this harb^{r} — does not allow us to proceed as rapidly as we are accustomed to do in sheltered waters. Got equal alts of sun on Friday and today for rates; I fear the firing at Rupert has upset some of my chronometer as I find a strange alteration in rate which I have never experienced before.

BULL HARB^{R} GOLETAS CHANNEL

AUGUST 27, MONDAY. Weighed at 6 am for Bull Harbour. Sent Shark to sound One tree Channel & rejoin us by the west of the I^{d}. The captive beauty "Hoo-saw-eye" went in here in hopes of meeting the "Otter" on her way to Nanaimo. At 7^{h}. 30^{m} entered Bull Harb^{r} and moored in 5 fms. A very snug little cove perfectly landlocked but entrance very narrow and a turn to make which a less handy vessel might refuse to perform; there is however a straight passage in — but very narrow & until we had examined it I did not feel justified in taking it. Cut a trail across from head of harb^{r} to North side of I^{d} not above 300 yards; we shall have to do the north side by walking as there is too much surf everywhere to land. A gloomy day SEly wind & some rain. I suspect there is always a swell on the Coast and I^{ds} outside this, for we have had very fine weather during the past week and no landing possible anywhere.

SEPTEMBER 1, SATURDAY. During the past week we have been much delayed by thick weather and surf; we got the north side of this Island in

[52] The name for yellow cypress among the Quatsino and Tlatlasikwala is *di'wał* (Pasco 1998: 15).

by walking round the coast at low water a long process owing to the rugged nature of the Coast. Succeeded also in getting some lines of direction to the Scott and Triangle Islands tho not very distinctly seen; a tide party observing at the small I^{d} at the entrance — the rise* [Richards' footnote in which he questions "as to the difference of tide between Bull Harb^{r} and outside"] and fall at springs not exceeding 12 feet which is strange as the outside coast close to shows a rise of very much more 16 to 18. The time of high & low water is regular. The Newhitti Indians[53] have just arrived and pitched their tents on the island for salmon fishing. They say during this month the salmon will be very abundant in the harbour. One of the Indians says there is plenty of coal on the coast of Vancouver I^{d} opposite — more than at Fort Rupert — which is probably true but there being no harbour near and a most exposed coast it would be comparatively useless. The north end of Vancouver I^{d} at any rate near the coast is high, rising in very remarkable hummocky hills (like waves rounded on their summits) and densely wooded.

The Shark returned on Wednesday having fortunately met the "Otter" and discharged Hoo-saw-eye to her — a most satisfactory riddance. This harbour has been surveyed and as soon as we get observations for the chronometers we shall start westward. The Shark was dispatched today with a month's provisions to return by the Eastern Channels working by the way. The Gig with Mr Mayne will follow her on Monday.

SEPTEMBER 2, SUNDAY. A gloomy day but succeeded in getting observations for rates.

BULL HARB^{R} AND SCOTT ISLANDS

SEPTEMBER 3, MONDAY. Unmoored early, weather thick with a good deal of rain. Gig left with Mr Mayne at 7^{h} am for the Eastward. The

[53] The Newitti, known today as the Tlatlasikwala First Nation, traditionally comprised three groups. The Tlatlasikwala occupied the eastern part of a territory which extended from Shushartie Bay, including Hope and Nigei islands, to Cape Scott and extending south to Sea Otter Cove. The Nakomgilisala occupied the western part of this territory. The third group, the Yutlinuk, occupied the offshore Scott and Triangle islands. By 1860, the three groups had begun to merge in response to declining numbers, disease, intertribal warfare and naval raids (Galois 1994: 291–294).

ship weighed at the same time and passed out of the Eastern channel, then shaped a mid-channel course to the west^{d}. About a mile west of Bull Harb^{r} or ~~from~~ at the west end of the Island of Galiano — commences the Newitti Bar — a Shoal bank extending across to Vancouver I^{d}. We carried 9 fathoms of stones for one or two miles when it gradually deepened to 13. 16. 18. 20. & 25 fms and this water we carried as far west as Triangle I^{d} the westernmost of the group off Cape Scott. About 40 miles immediately on getting out of Goletas Channel we found the flood setting to the Northward and we were set 4 or 5 miles north of our seeking by noon; the Ebb made about 1 pm and carried us southward again; so strongly did it set between Lanz and Triangle I^{ds} that the steam failing us about 4 pm (owing to Tubes being choked) it was as much as we could do to clear the west end of Triangle I^{d} off which a reef extends for some distance — a considerable westerly swell also helped to stop us. By 6 pm we had got an offing and bore up under sail South — the depth of water west^{d} of Triangle I^{d} having increased to 60 fms. 4 miles off and over 100 at 6 miles. This is rather a singular group extending in a chain westward of the north point of Vancouver I^{d}. It consists of 3 large I^{ds} and several smaller ones, high conical bare nine pin-shaped rocks; Cox and Lanz I^{ds} are the two easternmost ones. They are of about equal size wooded and from 800 to 1000 feet high. They lie 2 or 3 miles west of Cape Scott between which there appears a good clear passage. There does not appear a deep channel between Cox and Lanz I^{ds}; west of Lanz I^{d} extends a rocky chain of I^{ds} or rather bare rocks and lastly is Triangle I^{d}, the westernmost of the group which is a remarkable bare I^{d} about 1000 feet high — with a peculiar Notch summit. Between this and the rocky chain east^{d} there appear to be two good passages but until they are sounded they should not be used; after getting 15 miles south of Triangle I^{d} we steered in for the land intending to make San Josef Bay by daylight, and enter Sea Otter Harbour.

§ SCOTT ISLANDS AND QUATSINO SOUND

It has been observed that the flood was found to set to the north^{d} as soon as the ship passed out of Goletas Channel, and the Ebb to the South^{d}; now within Goletas Channel and as far south as Cape Mudge or nearly to Cape Lazo the flood comes from the north^{d} and ebb from the south^{d}, where they are again met by the Fuca Strait tides — the flood entering from the south[d] and meeting that from the north[d] about the Cape just mentioned. The flood tide therefore which comes from the northwest in Discovery Channel and Johnstone Strait & Goletas Channel is nothing more than the great tidal wave coming from the south[d] along the outside of Vancouver I[d] and entering by Goletas Channel to the north[d] as it does by Fuca Strait to the South[d].

SEPTEMBER 4, TUESDAY. At daylight we saw the land of Vancouver and steered in as we hoped for Josef Bay, but on approaching the land we found we must have been carried to the South[d], for nothing appeared answering to the opening on the chart. 57 fathoms was struck 5 miles off which soon decreased to 22 and we carried that depth to the entrance of an opening answering in some sort to Raft Cove of the Chart. As my particular object was to get to some anchorage where I could secure observations for the Longitude rather than to fetch any particular harbour on a coast all parts of which are equally unknown, I steamed along shore to the SE for a deep opening answering to Quatsinough off which we arrived at 10[h]. 30 am — having steamed along the land dist about 2 miles with 22 fms (gravel stones). Saw several canoes fishing for halibut off the entrance; a considerable swell existed at the entrance of Quatsinough and we saw several breaks — steered cautiously in to the NE[d] as we entered, the water deepened to 50 & 60 fms and within again we shoaled to 16 fms (rock) apparently on a ledge extending across the Sound. Passed a bay on the northern shore at the entrance, which seemed to offer good shelter but an Indian telling us there were sunken rocks, we kept on up the Sound. At about 6 miles up on the southern side saw a large village[54] and anchored off it in 10 fms, but finding 7 feet at a few yards from us weighed again and crossed over to the north side where after some search we found a very convenient harb[r][55] and running in moored in 10 fms. Got PM observations and were soon visited by the

[54] Likely Maate or Maylatee, now IR#8 Mah-te-nicht at Koskimo Bay (Galois 1994: 372). Boas records this place as *mEla'de* meaning "having sockeye" (Boas 1934: map 4–50). Richards' Survey Chart of Quatsino Sound shows three large rectangular houses at "Koskeemo Village" (Great Britain Admiralty 1860).

[55] Likely Koprino Harbour on the north side of the entrance to Quatsino Sound.

whole of the Indians belonging to the village off which we had first anchored.[56]

> SEPTEMBER 3. The Indians who inhabit the Inlet are divided into two tribes: namely Koskeemo and Quatsinough — the former by far the more numerous; and both are very wild, savage looking Chaps; unlike any other tribes we have met, they allow their beard and mustache to grow which adds considerably to the wildness of their appearance and are perfectly naked. The shape of the heads of these Savages is a Most Curious looking article — when young a bandage being passed tightly round it above the forehead compressing that part so much that the greater portion of the skin is squeezed into the Crown and they take the form of a most perfect sugarloaf — we measured around but one of the most remarkable was a little girl of 12 years old and from her eyebrow to apex of her Cranium was 8 in. laid off on paper with the proper Angles etc. it was a most remarkable object. . . . We traded for Salmon, deer etc. for a mere nothing — they appear not to know the value of anything — and for a small bit of soap could obtain 3 or 4 fine Salmon weighing from 12 to 14 lbs Each (Gowlland 1859–1864).
>
> SEPTEMBER 4. The Indians are named after the Inlet — very wild, savage looking chaps naked as the day they were born. Women ditto — except a clout round the waist. . . . They are very civil people but are a long way behind all the other tribes in civilization — they were astonished at our using matches to light pipes and Evidently had never seen any before" (Gowlland 1860–1863).

---

[56] The Koskimo were one of five tribes known today as the Quatsino First Nation: the Koskimo, the Quatsino, the Giopino, the Klaskino and the Hoyalas. Their combined territories extended from the Cape Scott area south to include Quatsino, Klaskish and Klaskino sounds. Richards and Gowlland identify only three groups: the Koskimo, the Quatsino and the Klaskino, although some Giopino still inhabited Koprino Harbour. By 1860, both the Quatsino and Giopino lived with the Koskimo at their principal village site at Hwates (Quattishe). The Klaskino were soon to join them there (Galois 1994: 347, 356–357, 363–364, 367).

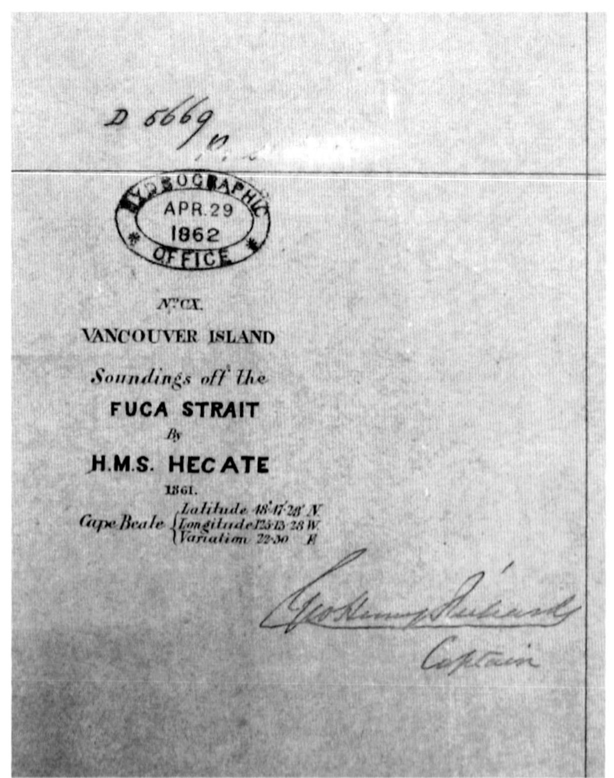

◀ Captain Richards' signature on an original chart from the H.M.S. *Hecate*, 1861. (UNITED KINGDOM HYDROGRAPHIC OFFICE: D-5669 FX#110)

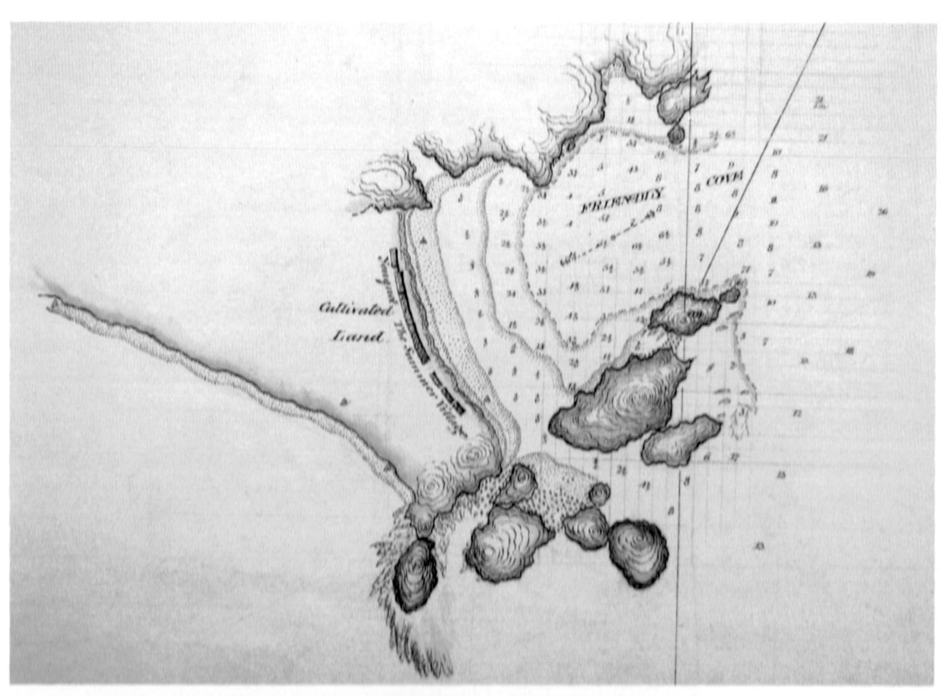

▲ Friendly Cove, Nootka Sound, Vancouver Island. (UKHO: D-6848 41A)

▲ Part of Discovery Passage by the Officers of the *Plumper*, 1860. (UKHO: D-6825)

DISCOVERY PASSAGE
S L A N D

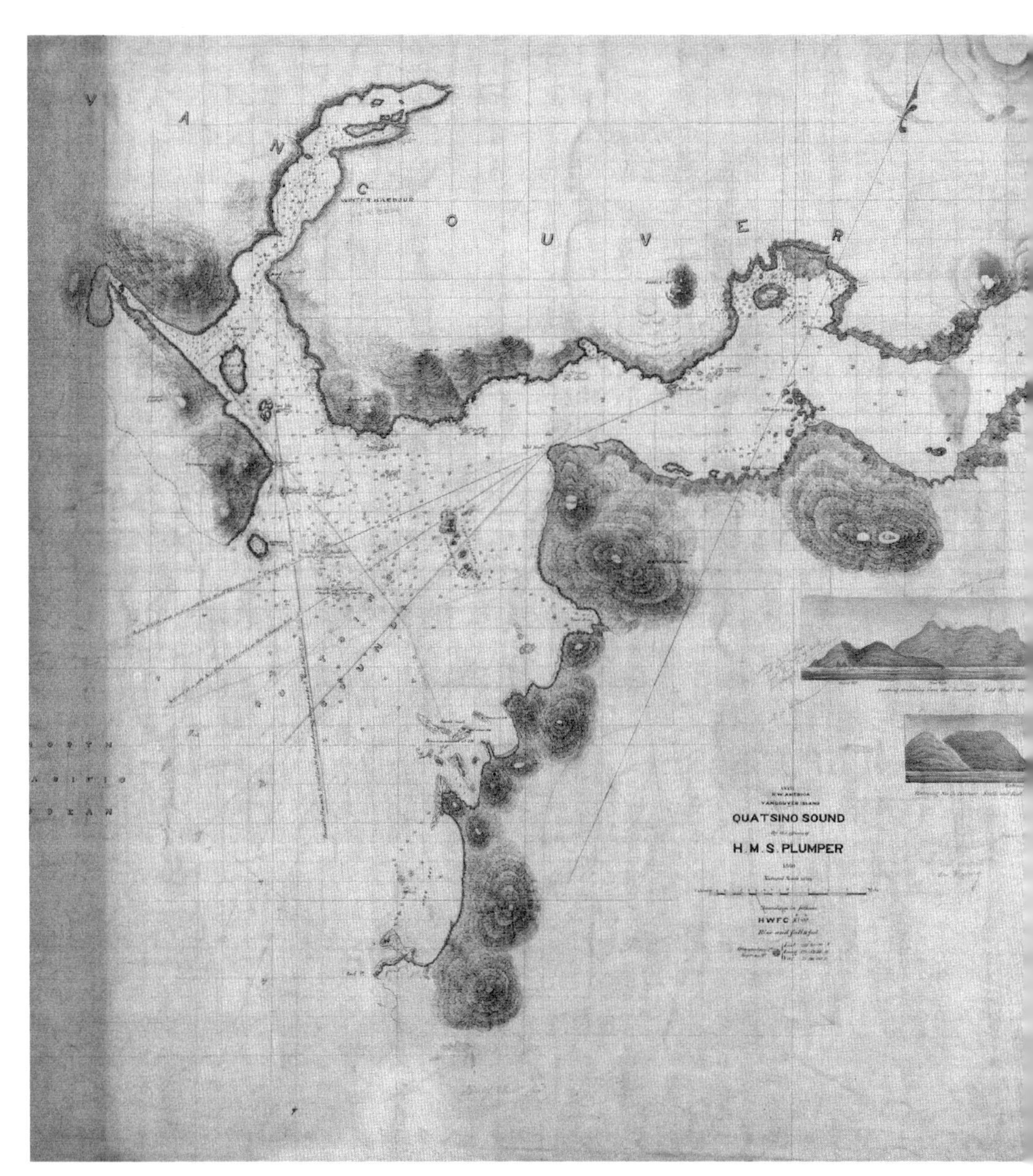

▲ Quatsino Sound, 1860. As was frequently his custom, Richards included insets of the landscape. A village with houses is shown at Koskeemo Bay.

(UKHO: D-6833 PREP 41A)

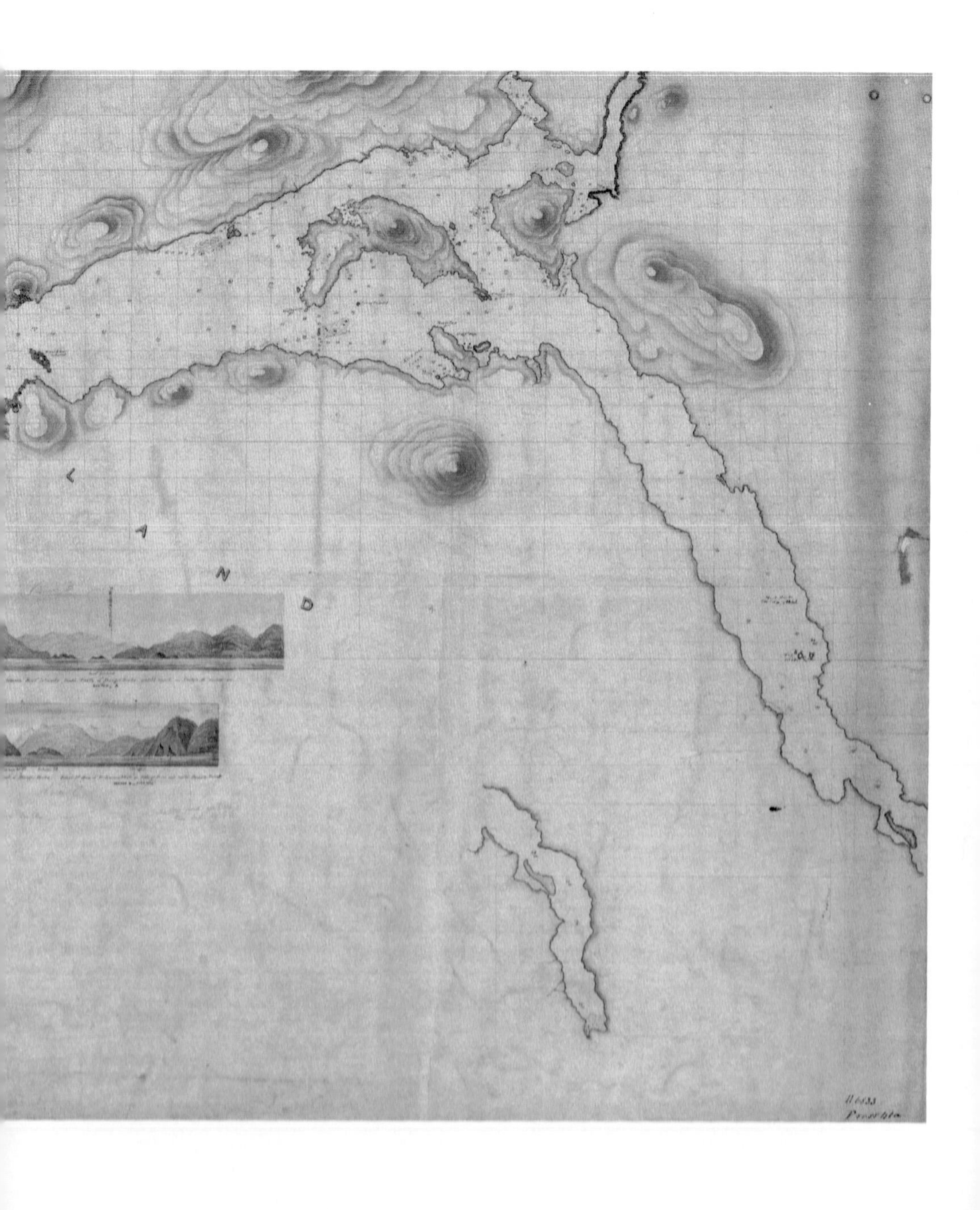
L
A
N
D

▲ Barclay Sound, 1861, showing Ucluelet Inlet. (UKHO: D-6700 ZA PART 8). The inset at bottom is a sketch of Ucluelet Village, 1859, by A. Bull on H.M.S. *Satellite*. (UKHO: D-6700 8N, SHEET 24).

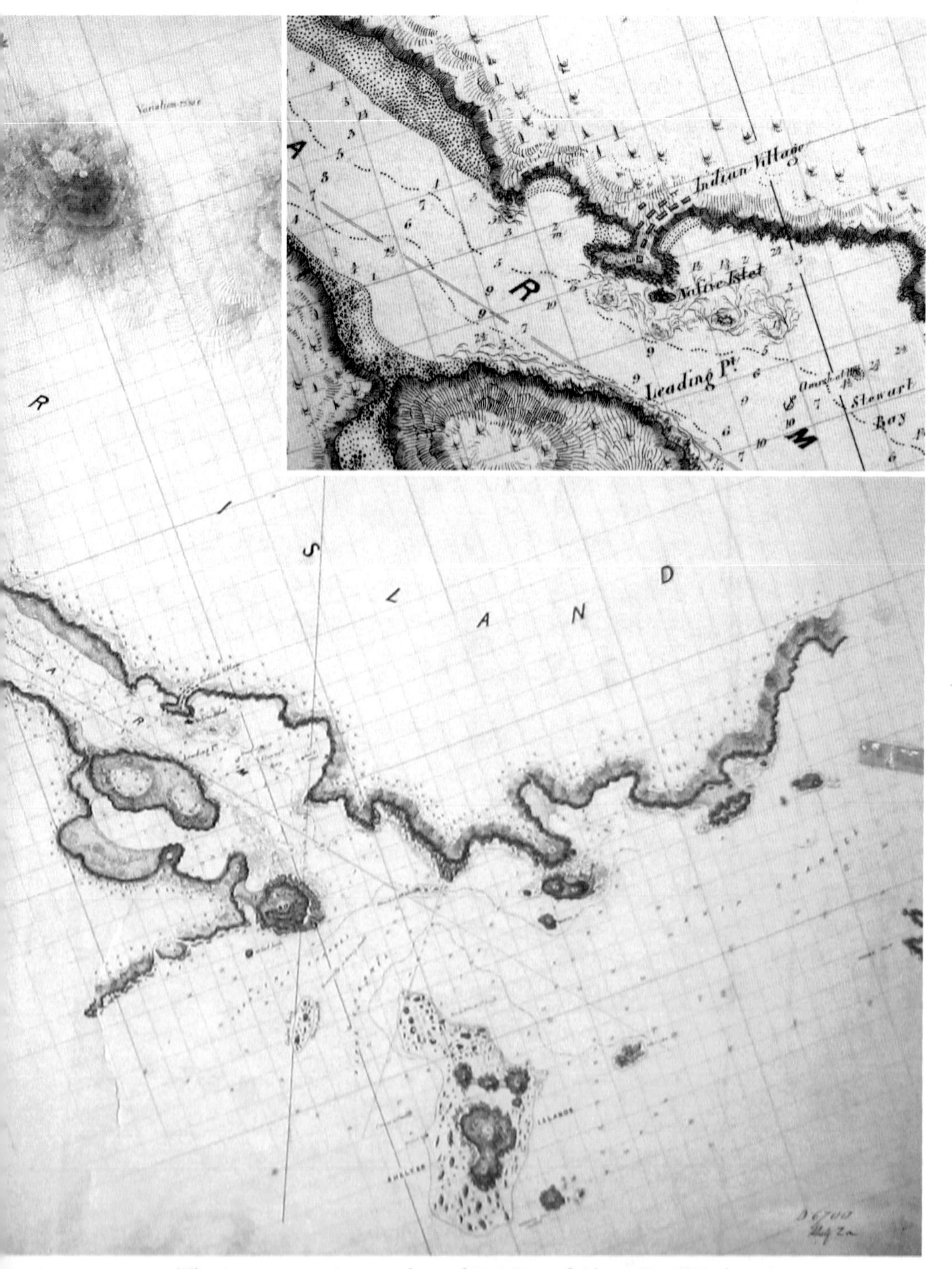

The inset at top is an enlarged portion of Chart D-6700 showing the houses of the "Indian Village."

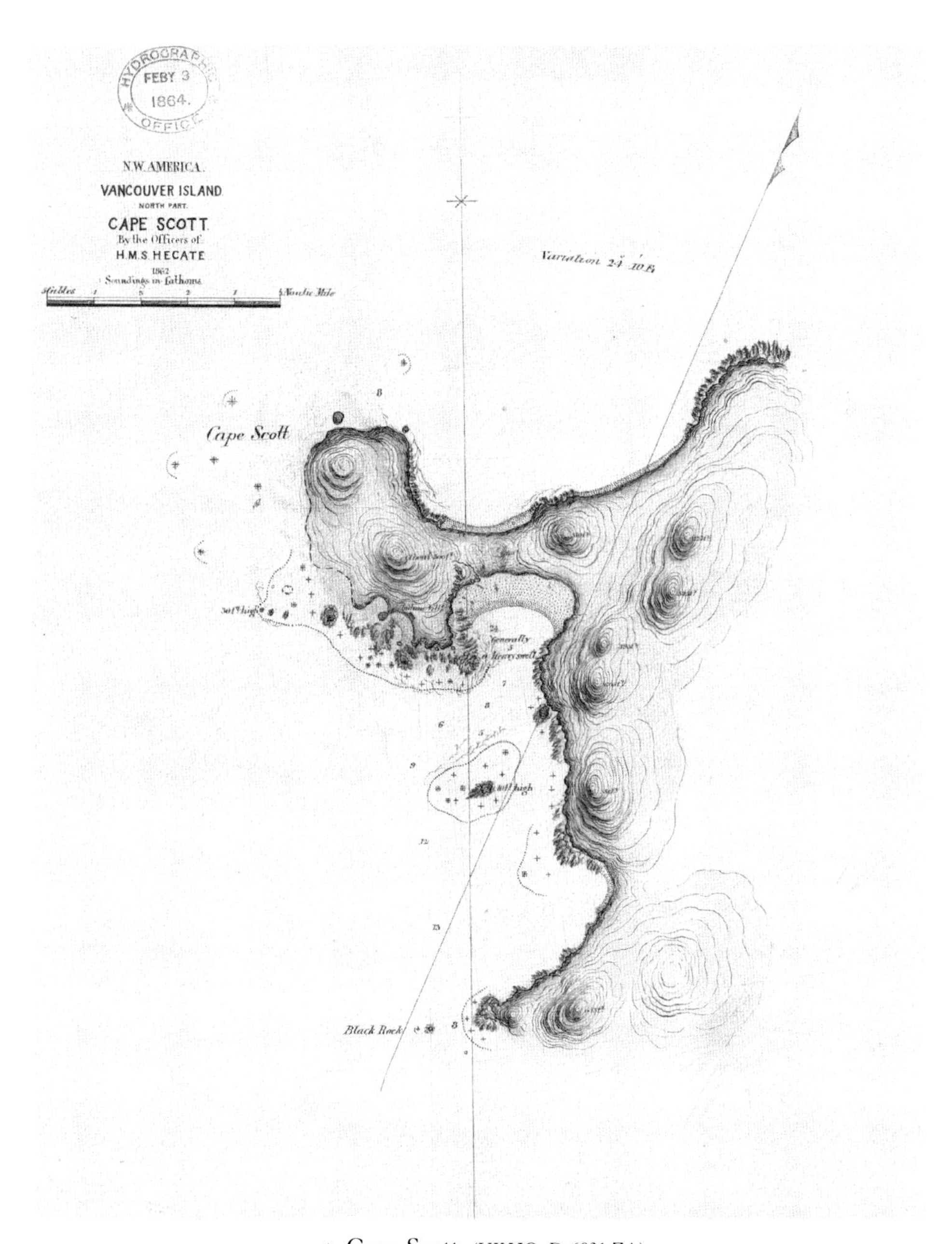

▲ Cape Scott. (UKHO: D-6831 ZA)

SEPTEMBER 5, WEDNESDAY. Sent the 2 whalers to survey the Entrance of the Sound while I got observations and prepared for the examination of the upper parts which are said to run within a short distance of Rupert on the East Side and for which place the Paymaster and a party started on Thursday morning the 6th.

> SEPTEMBER 5, WEDNESDAY. Whalers Started for the Entrance with 4 days provisions to commence the Survey — from the head of this Inlet which they say runs for 17 miles up. Port Rupert is distant only about 10 or 12 miles and Brown and Brockton started on Thursday on 5 days leave to go across (Gowlland 1860–1863).

SEPTEMBER 7, FRIDAY. I went in my boat accompanied by Dr Wood & Mr Pender to ascertain the extent of the Inlet in an easterly direction, and returned on Saturday eve the 8th finding it a far more extensive place than I had anticipated and more than we shall be able to finish with all our force during the next week. We passed several native villages which prove that the inhabitants must have been more numerous than they are at present or else that they are very migratory.[57] Many of the villages were deserted. My estimate of the number of natives in this inlet is about 400. They are ill clad and in a far nearer approach to a state of nature than any we have hitherto seen. There are a greater number of young people and children among them also, generally extremely well looking. Their heads are distorted in the most peculiar way; instead of being flattened as in other places they are formed when young into perfect cones.

[57] The Upper Quatsino Sound area, including Rupert Arm, was the territory of the Hoyalas before their assimilation with or conquest by the Koskimo prior to the European contact period circa 1770–1790 (Galois 1994: 361). It is possible that the village sites noted by Richards were remnants of Hoyalas villages. Historical population estimates also show that between 1835 and 1860 the Quatsino Sound people lost half of their population and after the smallpox epidemic of 1862 they were at 30% of their 1835 population (Galois 1994: 355). The timing of Richards' visit suggests another reason for the villages to be "deserted." Early September is salmon procurement season and the inhabitants of the villages may have been away fishing at the rivers near the heads of the inlets.

I saw some among the young girls extremely remarkable and we took the measurement of on[e] which were (see sketch).[58] This is accomplished by tying a bandage round the head of the infant and keeping it tightened up — the head is certainly more ornamental than the flattening and the process far more easy.

I tried to buy this girl with the intention of taking her to Esquimalt and getting her civilized — bringing her back next year, but neither the girl nor her parents would hear of it. The poor child, (she is about 10 years old) cried bitterly and I have rarely seen more feeling exhibited even among civilized beings; we gave her some presents of beads & earrings to pacify her — but she would not come onb[d] that day nor could we persuade her until the next to be measured when she only entered over the gangway, and then with her mother to whom she clung evidently in fear of being forcibly carried off — poor creatures they seem sadly in want of clothes and one cannot help regretting that so near a Christian community as they are, they should be allowed to remain in so utter a state of darkness and destitution. I fancy they had never seen a ship here before — larger than one of the small schooners which sometimes come for skins. They asked very high prices for the most trifling articles such as bows & arrows but it is scarcely to be wondered at, the desire to have clothes was very great and natural. I wish we could have provided all of them. They certainly appear to be the most good natured, obliging and harmless set of people I have ever met. They visited us daily — in their canoes sometimes 20 or 30 — and sat in them from morning till night gazing on the ship and getting I am sorry to say very little — but little we had to give them, but few of them came on board. The boats from the entrance returned this eve[g] having done a good deal of work — & discovered good anchorage in the bay at the entrance on the north side — with a channel leading from it apparently a long distance which there was not time to explore but where the ship must go as soon as we have got thro some of the work above. The weather up to this time has been unfavourable but little wind, but fogs and rain during the greater part of every day, which delays us very materially.

---

[58] Richards' journal did not include this sketch.

SEPTEMBER 7. Indians bringing us fresh salmon every day from their Salmon weir which is most simply and ingeniously formed of a number of stakes driven into the ground at L.W so as to be above H.W mark encircling a larger part of the beach, having a narrow door in the stakes Seaward for them to Enter. When once in they can never escape again and continue there till the tide falls when the indians come and Spear them (Gowlland 1859–1864).[59]

SEPTEMBER 9, SUNDAY. A foggy morning but cleared so that I was able to get observations; no natives this morning but great numbers in the afternoon — they evidently have some idea of the Sabbath for they were cleaner than usual as far as apparel went and decorated with feathers (goose down) stuck over their heads. They went thro all the Roman Catholic jugglery, chanting the hymns taught by the priests on the east side. They have not I believe been visited by them — but one of the Indian Converts from Rupert has been among them[60] — does this reflect on us, I fear so; the State ought to depart from its non-intervention system in religious matters in such cases as these — and at least the moral teaching, a primary civilizing operation, ought to be undertaken or paid for by the Gov^t to whom the Colony owes allegiance — if the Colony can't afford it themselves.

SEPTEMBER 8 [9], SUNDAY. The Indians paid us a visit in procession as if going to Mass, consisting of the tribe both Men women

---

[59] Barrett-Lennard (1862: 81–82) noted a similar wooden stake fish trap in 1862 when he visited this area. He noted that at the mouth of a river that flowed "into the south of Koshkeemo inlet," they had to "steer clear of a number of stakes; these being interlaced with slips of bark, formed a salmon weir, which, while affording ingress to the salmon at one particular spot, prevent their finding their way out, unless they happen to strike that same spot again."

[60] In 1863, the Roman Catholic mission withdrew from Fort Rupert as the "Indians there, with too few exceptions, remained callous to divine teaching" (Theodore 1939: 231). In 1878, Rev. A.J. Hall opened an Anglican mission at Fort Rupert (Fisher 1992: 137).

> and Children numbering between 100 and 150 — one Man who seemed to be priest amongst them stood up and led off the choir chanting in their own language the Roman Catholic church Service — and they certainly sang very well indeed (Gowlland 1859–1864).

SEPTEMBER 10, MONDAY. The two Whalers left this morning with Mess[r] Bull and Pender to survey; the former — the East arm leading to the Rupert Portage[61] where I had gone on Friday last and the latter to trace an extensive arm running in a southerly direction[62] from Limestone I[d] [Drake Island] 8 miles above this and which I think likely communicates with the sea.

I should have observed before that Limestone was found in large quantities on this Island, perhaps 14 miles from the Entrance of the Inlet — and also for a considerable distance above. Limestone formation occurred over a large space — very easy of access. Specimens of coal were brought from Quatsino bay at the entrance by M[r] Bedwell — and there can be no doubt but that the coal measures exist here. If this is established it will render this port one of great importance — for it is proved to be perfectly easy of egress for ships of any size sailing or steaming — with many good anchorages. I was employed myself today in surveying the Inlet in the neighbourhood of the anchorage and connecting it with the work of the parties — Eastward and Westward. The weather was rainy until 10 am then fine for the remainder of the day until sunset — but at the upper end of the Inlet I hear they had much rain. The winds during our stay have been variable and very moderate — fogs every morning until 9 am — all day cloudy, as a rule rain some time during the day — generally all the afternoon — but when the sun appears very scorching. I fancy the regular summer winds — westerly — from SW

---

[61] Richards notes the presence of a trail here in the *Vancouver Island Pilot*: "From Rupert arm to Hardy bay, on the north-east side of Vancouver island, is a distance of only 6 miles, and a trail exists between the two places, much frequented by the natives for trading purposes to Fort Rupert" (Richards 1864: 249).

[62] Neroutsos Inlet.

to NW cease about the end of August, and that during September this calm foggy, rainy, uncertain weather occurs as a prelude to the real setting in of winter. The summer throughout the region has been a very wet and uncertain season. Nature seems to have taken a freak. The berries have not ripened — at least the Sallall[63] the principal one the natives depend on. The bright red berry,[64] a species of cranberry, on the contrary have flourished wonderfully — they like wet while the Salall cannot bear it.

SEPTEMBER 11, TUESDAY. Surveying all day in my boat. Visited the native village on the south side[65] — a very filthy place — but almost deserted as the inhabitants line alongside the ship all day. Paymaster & his party returned from Fort Rupert this morning, having had fatiguing work; trail from head of E^{n} arm about 12 miles nearly due north & not good walking; they brought several bags of potatoes back with them — owing to the kindness of Capt. Mitchell HBCo in charge at Rupert. Rain came on at 2 pm as it almost invariably does in the afternoon.

> SEPTEMBER 11. Mr. Brown (paymaster) and Mr. Brockton returned from Fort Rupert with the Indians bringing over a good supply of potatoes and vegetables generally; which were very acceptable as these Indians unlike a great many of their neighbours have no Idea of Cultivating the land — all they care about is to get fish Enough to feed themselves and have no further thought of Cleanliness or Comfort, their blankets are made from the bark of trees which they Manufacture Most ingeniously into a very warm covering which answers the purpose of a blanket admirably (Gowlland 1859–1864). They reached it the same day of Starting very tired having Estimated the distance at about 12 miles from the head of this Inlet (which runs

---

[63] Salal (*Gaultheria shallon*).

[64] Possibly the bog cranberry (*Oxycoccus oxycoccus*) or the red huckleberry (*Vaccinium parvifolium*).

[65] Maate at Koskimo Bay.

some 16 from the Anchorage) thro a very bad Indian trail in a NW direction and accomplish the distance in six hours (Gowland 1860–1863).

SEPTEMBER 12, WEDNESDAY. Today we had a fresh SW wind which brought very fine weather and entirely cleared away the foggy atmosphere which has been hanging over the land for the last fortnight. Galley and Dingy employed surveying this anchorage and sounding it. I ascertained from the Chief of the village opposite thro our interpreter, Jim from Rupert, that the male adult population in the Inlet amounted to 137 — the number of women and children he promises to get me tomorrow. I should imagine they must amount in all to nearly 500.

SEPTEMBER 16, SUNDAY. The two whalers returned last night — the SE arm having been completed but not the eastern one leading to the Rupert portage. The weather has been most unfavourable. Rain on every day but one for the eleven days we have been here. As there is another extensive arm running to the NW which the natives say leads to within a few hundred yards of Newhitti on the shores of Goletas Channel — and which I cannot finish this season[66] I shall probably leave the unfinished eastern arm with it until next year — and go tomorrow to the entrance bay known as Quatsino — from which an inlet also runs to the north[d] — there are some reefs to be examined yet in the entrance, and I am anxious to visit Nooka and secure observations before the wet season permanently sets in. The barometer has been rising for 2 days and is now at the extraordinary height of 30.36. The day is fine — foggy till 11 am. Mr Brockton the chief engineer went to the entrance of the NW arm leading to Nawhitti last week to examine some coal — which the Indians gave information of — he found a seam about 2 feet thick at low water mark and procured 400 cwt[67] of very fair coal. Of course the specimens procured were the least favourable from the fact of their being

---

[66] Holberg Inlet runs westward and its head is approximately fourteen miles (twenty-three kilometres) from the mouth of the Nahwitti River. The full length of the Inlet was not surveyed until 1863 by Captain Pender.

[67] Hundredweight.

covered at h water but Mr B. does not consider there would be any difficulty in sinking a shaft within high water mark. The cook used this coal in Grant's Galley and it did all the requisite cooking and condensing of fresh water.

SEPTEMBER 19, WEDNESDAY. The weather remarkably fine — since Sunday barometer above 30.30 and a NE wind. On Monday 17th we left our snug cove and steamed up the arm sounding, passed round Limestone I^{d} and down on the So side of the Sound and anchored at 3 pm in the bay immediately within the entrance on the NW side which the natives call Quatsino.[68] It is a good shallow anchorage in 7 fathoms within the 2^{d} Island; dropped a boat outside to look from the breaking rock we saw on entering but she returned without finding it.

On Tuesday morning boat went at low water again to search for the entrance reef while I went with 2 boats to explore the arm running NE from this bay and which the natives informed us was very extensive.[69] I found however that it only ran 4 or 5 miles and terminated in a creek. There is good anchorage in this arm immediately within it and for 1½ miles above. The whaler returned in the evening having found and fixed the position of the breaking reef. In addition discovered another also breaking at low water — and on plotting our track in on our chart I find we passed a few yards only from this latter danger which if we had struck must in all probability have proved fatal to us, for there was a heavy swell at the time — on getting 12 fathoms we stopped and sent a boat ahead; at this moment the rock must have been almost grazing our port side which I consider a most providential and narrow escape. There is a passage in on either the north or south side of these rocks; that to the south the regular fairway — we passed between them in our ignorance of their existence — close south of the northernmost one. There is a village at Quoskimo[70] containing 60 inhabitants — including women & children, and the other parts of the Sound 460, making a total of 520

[68] Forward Inlet.

[69] Winter Harbour.

[70] Owiyekumi near Village Islet in Forward Inlet was the winter village of the Quatsino for most of the nineteenth century (Galois 1994: 375–376). Now IR#11 O-ya-kum-la.

souls. We found them a mild and harmless people — obliging and no case of dishonesty.

More children among them than I have seen elsewhere. They were almost the only natives I think we have met who exhibited any signs of astonishment at our propeller and they were evidently surprised at it.

SEPTEMBER 20, THURSDAY. This morning at 5.30 we left having obtained observations yesterday for chronometer distances. Passed out north of the reefs between them & NW Entrance I^{d} — a strong NE wind with a heavy bank of fog rolling out of the Inlet. Shortly after getting an offing heavy breakers were reported to the SW. They turned out to be many large whales disporting in the sea, and throwing up immense columns of water certainly very much resembling breakers. Indeed many could not be considered that they were not for a long time. Steered to pass Woody P^{t} 4 miles. The wind drew from NW and freshened so much that we could not get soundings without rounding to every time. The bank which extends off the northwest point of the I^{d} appears to cease off Quatsino — for we got no bottom with 150 fms — when approaching Woody P^{t}. This pt is a bold promontory with a remarkable double Islet lying off it. It is very badly named altho by Capt Cook. First, it ought to be a Cape and not a point, and next all the coast is densely wooded equally so with Woody Point. I shall call it Cape Cook. At 9.30 am raised the propeller and made sail in the hope of reaching Nootka tonight and saving observations tomorrow as well as a night at sea. After passing south of Cape Cook or Woody Point the high central mountains rising over the Nimpkish Valley come into view — they are now almost bare of snow. NW^{d} of Woody Pt is Claskimo, a place for boats or schooners — so says our Interpreter Rupert Jim. Immediately SE of the pt is Nasparti which he describes as good for ships and plenty of sea otter to be had, which means 3 or 4, for the natives set such a value on the otter skin that the neighboring tribes always know how many there are at every place ready for sale. Jim could tell us at Rupert how many there are at Nootka.

A rather deep bay or indentation on the Coast access between Cape Cook (Woody P^{t}) and the port named Esperanza on the chart; in this bay there appear to be several ports and Jim says that in one of them,

Kai-o-quot [Kyuquot Sound] there are a large number of Indians — he says that there are 300 fighting men[71] but does not know how many including women and children. From our experience there must be at least 700 in all, taking the proportion of women and children at Rupert & Quatsino. They were far more numerous a few years since at Kai-o-quot but the Nootka Tribe who are on bad terms with them have killed great numbers — according to Jim. Shortly after noon today the wind moderated and at 3 we got steam up. Passed the carcase of a dead whale within a mile which was reported as a rock and appeared very much like one. The numerous whales seen today would be extremely liable to give rise to reports of reefs for we were almost deceived at the moderate distance we saw them so considerable was the apparent breaker caused by them — I saw one jump right out of the water certainly 50 feet long and as he fell the sea presented the appearance of a break over a sunken rock to all intents. If I had not seen the fish out of the water at the moment but only seen the break of landing I should most certainly have gone to search for the rock — which I should have believed to exist. We saw some very remarkable mountains today — but so many and such a sameness in appearance that they almost confused me — some I certainly recognized as old friends seen on the East coast but up to this time I have not been able to identify the real Conuma peak — for a positive certainty. At Sunset — 6 pm we were still some 12 miles from the Entrance of Nootka — but the weather being fine and trusting to some slight recollection of the Land on my visit there with "Sulphur" more than 20 years ago[72] we went on — in the hope of fetching 'Friendly Cove' [Yuquot]. We struck soundings with 25 fathoms some 6 miles offshore at least 5, and found we shoaled off the low land which terminates at Maquinna point — kept off a little — and deepened to 26 fms — picked our way in by the lead. The water deepening after passing Maquinna pt. Steamed along the shore for Zucutan pt and although pretty dark by this time we shortly opened the little cove scarcely a cables length wide & entered making a running moor; the cove is extremely small and I was almost

[71] A footnote dated June 1862 reads: "400 fighting men and 1600 all told — as gathered by Mr Gowlland 1862."

[72] Richards' illegible footnote.

inclined to think I had mistaken it so very small did it appear by the almost darkness.

SEPT^{R} NOOTKA SOUND 1860

SEPTEMBER 21, FRIDAY. This morning I found we were in as good a berth as if we had picked it up by daylight — so we put the mooring swivel on and parties landed on Sulphur's OB I^{d}, the outermost of the rocky I^{ds} on the SE side of the cove for observation astronomical and magnetic. I received sightings for time and Latitude very satisfactorily, but as we had completed the magnetic observations we discovered a difference of 12 degrees in magnetic bearing between 2 spots on the I^{d} not 50 yards asunder thus proving that the needles were affected by some influence in the rock. The Islet is bare of trees, some 40 feet high and of Trap rock with appearance of Iron. A piece of the stone applied as a magnet to the needle did not affect it. Yet strange to say the difference of 12 degrees existed by several compasses. After finishing my afternoon observations, I left my sextant, Horizon and the pocket chronometer on the spot covered with a canvas screen, while I got a sample of angles about 10 yards distant on the summit of the Islet. While getting my angles 2 natives landed in a Canoe and came to us — explaining that they wanted to get their musket repaired. I told them to take it to the ship, intending when I went off to cause the armourer to put it to rights. They left us with some satisfaction to me for their scent was extremely strong and I fully believed they had gone to the ship. I remained on the spot for an hour, waiting for Mr Browning to fix a micrometer levant, which I wished to measure and which I had sent him in my boat to put up before taking the chronometer onb^{d} to compare, contrary to my usual practice — but I had another set of Equals coming in which, as the sun was getting clouded appeared doubtful. Still I thought it better to keep the watch on the chance of getting them, & Mr Pender was present to take time. The sun remained obscured & I did not go to the OB to try for them but remained at the theodolite having finished the angles. Mr Pender & myself went to pick up the instruments and carry them to the Galley just approaching with Mr Browning — my surprise and horror may be imagined on finding the Chronometer gone — it was in a box case and a fearnought Cover over all. I hailed the ship immediately for the 2 whalers and sent the Galley & 2^{d} whaler in chase of the canoe with our two friends which was at the moment seen entering a labyrinth of I^{ds} 3 miles off. I felt no doubt they must be the culprits. At 7 the Gally returned having lost all trace of the canoe. At 7.30 2^{d} Whaler returned with her in tow & the two thieves. They had seen her just entering a cove and the culprits finding no hope of escape landed, hung the Chronometer on a bush, pointed to it and concealed themselves in the bush. However by the aid of our interpreter Jim one was secured and a canoe passing at the moment with 2 Indians Jim demanded their assistance and the other fellow was taken and both brought on board. I caused them to be put in Irons and the next morning dispatched a native for the Chief Maquinna to say I should be glad to see him — and to tell him I had known his mother's father the Old Maquinna 20 years ago.[73]

On making enquiries from the people here thro Jim I found the Old Chief who had existed in the Sulphurs time had been long dead — that the son mentioned by Cap Belcher was also dead as well as another one we did not then see; that the little black-eyed good looking girl also mentioned by him and then about 13 years old was alive and married and that her son was now the Chief with the name of Maquinna, or the grandson of Old Maquinna on his daughter's side. There were only 2 natives at Friendly Cove when we arrived, the whole population being at Tasis salmon catching and Curing — but on Saturday they began to return having heard of our arrival. I was very anxious to see Maquinna — and still more so his mother, now described as the old woman but who I well remember 21 years ago as a pretty little, almost white girl.[74]

---

[73] On October 3, 1837, Richards was a young midshipman on the *Sulphur* when he met the "Old Maquinna" and his family at Nootka Sound. The *Sulphur* was a survey ship under the command of Captain Edward Belcher sent to the Northwest Coast to examine the mouth of the Columbia River and San Francisco Bay (Gough 1971: 43).

[74] In 1837 Belcher wrote: "The young lady here introduced was yet but a child. Her features were, however more of the Chinese or Tartar breed, than those of the brother. Her manner was very simple and winning; she had black expressive eyes; and her affection for her father, on whom she often clung, with her head reposed on his shoulder, was quite a novel sight amongst these people. The son, as well as the daughter, appeared to receive all the respect due to high rank, even from the father, who invariably turned over his presents to them . . ." (Belcher 2005: 109).

SEPTEMBER 22, SATURDAY. It rained constantly all night, and until 9 this morning when it suddenly cleared quite fine & I got all the observations of yesterday over again. For Dip & Variation we changed our quarters to the sandy village bank but on trying the Compasses again at the I^{d} of yesterday I found no discrepancy in the bearings which is very remarkable. There can have been no mistake for 3 different Compasses were used yesterday at several spots on the I^{d}. Perhaps we may throw some light on the business before we leave. About 50 natives arrived at the Cove today but not Maquinna. They have a few small sea otter skins but ask a most exhorbitant price for them — 20 blankets. They have also some small potatoes which they offer — 2 small baskets for a blanket. They must fancy us extremely raw. I landed at the Friendly cove village today. It is the skeleton of one of the largest villages I have seen any where. Nothing however but the upright and poles are standing; all the boarding and mats are removed I fancy to the upper places where they are taking salmon. There are 6 or 7 acres of cleared land round the village which occupies the site of the Spanish Settlement of Quadra.[75]

I visited the patch of land of Mr Meares which was the cause of so much diplomatic work between the Spanish & English Govts.[76] It is accurately described by Vancouver as a triangular patch enclosed between steep rugged rocks — no side of the triangle being 100 yards in length. The natives grow potatoes on it now.

Friendly Cove is a very small place indeed — just room for 2 ships of our class to lie moored with their own anchors. In Quadras & Vancouvers

---

[75] In 1789, the Spanish authorities dispatched three ships from San Blas to Nootka Sound where they built a fortified settlement at Friendly Cove, just inside Yuquot Point (Gough 2007: 118). They called it Puerto de la Santa Cruz de Nuca but it was also known as Sala de Los Amigos, the Spanish translation of the English name, Friendly Cove (Hayes 1999: 68–69).

[76] Richards is referring to the Nootka Controversy, a series of incidents between 1789 and 1793 which nearly led to war between Britain and Spain. It was resolved diplomatically in three treaties known as the Nootka Conventions that eventually led to the mutual abandonment of Nootka Sound. (see Gough 2007: 118–127; 267 and Nokes 1998: 123–169). Soon after the departure of the British and Spanish, the Mowachaht reclaimed the site: the "remnants of the Spanish buildings had been replaced by Maquinna's lodges . . . and Friendly Cove took on the appearance much like it had . . . when the Spanish . . . first anchored off the Sound" (Gough 1980: 131).

time six small craft seem to have been here together — but they were very small and must have been moored with hawsers to the shores. A swell generally rolls into the cove but nothing to make it uncomfortable and I should say it was always perfectly secure.

The Nootka Tribe I am assured by the people here numbers 457 — including all. Our interpreter Jim found this out from one of the natives who cut 457 sticks up like so many matches to explain the number. For a grown person a stick about 8 inches long and a smaller one for children, all of which he brought to me.

> SEPTEMBER 22, SATURDAY The Indians began to arrive in their Canoes, from Tasis a village some 20 miles up the sound . . . (Gowlland, 1859–1864).

SEPTEMBER 23, SUNDAY. Macquinna arrived today in a large Hyas[77] canoe — he brought his wife and two or 3 relatives with him, including his Uncle, brother of the old man in the Sulphurs time who perfectly remembered our ships and described the fire works and magic Lantern we exhibited for their amusement.[78] Maquinna is a lad about 19 or 20 years old — not the most intelligent I have met among the Indians — indeed rather stupid looking — his wife is about 16 or 17 I imagine, very like a seal — a good natured fat creature. I had them onboard and first made Maquinna give a severe lecture to the 2 Thieves still in Irons who stole the Chronometer. This he deputed his Uncle to do, who harangued them in a very ferocious and threatening spirit — occasionally striking them with his fists, and assuring them that if ever they stole again they should lose their heads. They appeared in a great state of alarm and looked very much ashamed of themselves. After receiving their lecture

[77] Chinook jargon for large, great or mighty.

[78] "At dusk I landed, taking with me a magic-lanthorn and supply of fireworks. At the former they all exhibited the most unfeigned delight, to a degree quite outrageous; but at the ascent of the rockets, their impressions amounted to fear . . . and by the light from the fire, I could perceive the tears rolling down the cheeks of Macquilla's wife and daughter, who fled to the Bush the instant the fireworks were over" (Belcher 2005: 110).

I told them they were at liberty to take their canoe and go — which they did in the shortest possible time — and no doubt they will attend to my injunction not to come near the ship again.

After this scene I made some presents to Macquinna & his family — some blue cloth Blankets, Shirts, Tobacco, Soap, Knives & axes — with beads and rings in addition to the ladies. We also regaled them with biscuit and weak wine & water. I was sorry to have no rice or Molasses which is the orthodox feast for these people; however, they did justice to the biscuit and the seal was very sleepy for ten minutes with the small quantity of wine she had taken. She soon however recovered. Our decks were crowded with them for the whole afternoon. They had little to dispose of but mats and the mantle made of the Cedar bark. My royal visitors I found as keen at trading as any of their subjects after they had stowed away their presents. The seal made an excellent bargain with my steward for some mats — getting no end of Tobacco in return. After they had retired Wicannish's wife came to ask for her share of presents as her husband was a Chief like Macquinna but away deer shooting. I accordingly treated her like those who had just departed — decorating her with rings and Earrings.

> SEPTEMBER 20 [23]. . . . we were visited by one solitary canoe with two men, and they remained about the Ship all day, until in the afternoon the Captain went on Shore to take his afternoon Sights and the Indians landed with him, watching his movement and expressing their astonishment as they always do at the Sight of the Instruments. Shortly after finishing his observations the indians left and made sail paddled away up the Sound. Some two hours afterwards the pocket Chronometer which had been laid down on a rock was found to be missing and suspicion immediately alighted on them. The galley went away immediately to chase and shortly after the 2nd Whaler. At 1:30 the Galley returned not having been successful in catching the culprits altho' she pulled some 8 miles up. About 1 hour after the 2nd Whaler returned having proved so fortu-

nate as to catch them — it appears they thought the Chronometer (which was wrapped up in a flannel bag) was tobacco or something of little value and took it but when the boat got close to them they suddenly ran the canoe into a small creek in the rocks and made off into the forest first hanging the watch on the bough of a tree and calling the attention of the crew to it. They were however pursued and taken, brought on board and snugly lodged in Irons under a Sentrys charge. Next day the Captain sent a canoe to the head for the Chief, Macinna (the grandson of the one mentioned by Captain Vancouver) to come down and bring a number of his subjects to witness the punishments of the defaulters and the following day he arrived accompanied by about 2 hundred indians and on Saturday 23rd of September, this Gentleman came on board accompanied by his wife (the Queen) — Nephew (a little dirty naked lout) and Uncle besides several others of the Royal family. They all smelt horribly of stale fish etc. but were otherwise rather clean; and taken into the Chart room where several sorts of presents were distributed to each member. The king of course getting the lions share, such as Blankets — Beads, Looking Glasses, Soap, tobacco, Knifes etc. at which the[y] were highly delighted. The prisoners were then brought aft, and at the Captains request the Chief was told to speak to them and set forth the Enormity of their Crime and the punishment awarded in H.M.S. for such offences and to blow them up well. The poor wretch appeared as much frightened as the Culprits and set his Uncle on, (who I suppose is the Court bully) Anyhow the old rascal began in a low tone and gradually worked himself into a towering passion pushing them, striking them repeatedly; and screeching out his words in a great state of excitement — the culprits all this while towing a line on the quarter deck and looking very penitent & frightened — and no wonder for this tedious old scoundrel would frighten old Nick by his outlandish gesticulations — after he had yelled himself out of breath, he stopped and appealed to Capt R. if that would do — and having been heartily tired of this tirade such a consent was given and the Indians

> pardoned — they were not long scuttling off to their houses after going over the side after which the people were all allowed on board to trade and sell their skins etc. (Gowlland 1860–1863).

SEPTEMBER 24, MONDAY. This morning I went with 2 boats to Tasis accompanied by Mr Pender & Dr Wood. We had a fresh Northerly wind against us all day and did not reach the village of Tasis at the head of the direct arm from here until 4.30 PM, Maquinna being below with his family and the greater part of the Tasis people. We found the village rather quiet & pitched our tents about 300 yards below & on the only available piece of land we could find not covered with trees & brush wood. The village is a large one but not so much so as the Friendly cove place. The houses very large & I think cleaner than others I have seen. The people were very civil helping to land our traps and carry them to the tents — nothing but Wakash[79] was to be heard constantly repeated by men, women and children. We set off some rockets in the evening but they did not express much astonishment or indeed curiosity and few except the men came out to see them. From Tasis to Nimpkish on the East side is a journey from 3 to 4 days, the great lake being only half a day from here. The weather was remarkably fine and we had an excellent view of all the peaks — also got latitude and time.

SEPTEMBER 25, TUESDAY. After breakfast & having got forenoon sights we started down, the whaler sounding for anchorage in the basin, but like the heads of all these inlets the deep water runs to their heads, rapidly shoaling from 30 fms to 2 or 3. We were favoured with a land wind down the arm for the first 12 miles and camped to lunch at the same spot as on going up on the west side just opposite a considerable village

---

[79] When Captain James Cook arrived at Nootka Sound and met the Mowachaht, he applied the term "Wak'ash," a Nootkan expression of friendship, to the people. He also erroneously named the inlet "Nootka" from the word *Nootk̲a'a* meaning "go round," a direction given to Cook to "go round" the point into the inlet (King 1999: 125). Linguists use "Wakashan" as the name of the language family that comprises the languages spoken by the Makah, Ditidaht, Nuu-chah-nulth, Kwak'waka'wakw, Laich-Kwil-Tach, Haisla, Oowekeeno and Heiltsuk people (Thompson and Kinkade 1990: 39). Each of these languages is divided into numerous dialects.

& salmon stakes. Here we got a latitude and proceeded down, shortly meeting the Southerly wind or Sea breeze, westerly outside but blowing right up the arm. At 3 PM we met Maquinna and his family returning to Tasis as they said, to see us. So they turned back and camped again at Friendly cove which we reached at 4.30 PM. The 2ᵈ Whaler had been employed surveying the lower part of the Inlet during our absence. Got star latitudes tonight at the village. Maquinna & his ladies came onboard and devoured some more biscuit which they seem very fond of.

SEPTEMBER 26, WEDNESDAY. The weather remarkably fine as indeed it has been almost without intermission for 10 days & very clear — a sea breeze blowing during day and land wind at night. It is beginning to feel cool now towards Evening. Wicannish came today & brought a Deer — which I was told he wished to give me. However I did not feel desirous of taking it from him and I gave him similar presents to those Macquinna had received. The interpreter asked me after if I wanted the Deer; if not Wicannish would sell it for some shirts to the people onbᵈ. I replied — certainly not; sell it by all means — at which he appeared wonderfully relieved. He is a more intelligent manly kind of fellow than young Macquinna, perhaps a little older. Macquinna seems slightly under petticoat government and I suspect the Seal is the best man of the two.

❧ SEPTᴿ NOOTKA SOUND–PORT SAN JUAN 1860

SEPTEMBER 27, THURSDAY. Unmoored at daylight and left Friendly cove at 6 am. Carried soundings varying from 20 to 30 fms 5 miles off shore, when off to 50 fms. Steered along the Coast 10 miles offshore sounding. Wind SEly with occasional intervals of fog. Baro. had gone down this morning & I expected a SE gale, having had so much fine weather. However at 9 am it rose again and we steamed all day against a fresh SE wind with fine weather sounding — the greatest depth 10 miles off 70 fathoms, generally much less. At a little after midnight saw C Flattery light and steered for Port San Juan where we anchored at 7 am. This port offers a good stopping place with SE winds — but is entirely open to those from SW when a considerable sea would roll in — and altho small vessels are protected by a jutting point on the south side of the bay near the head during moderate winds, yet it is said that during a heavy SW it breaks right across in 3 or 4 fathoms. The wind generally shifts suddenly from SE to SW. Therefore a vessel should not remain after the end of a SE gale. When the lights on Race Rocks & Fisgard I^{d} are established, steamers would never require to stop at San Juan inward bound — outward bound perhaps. Neeah bay offers more convenient anchorage with a SE wind. The San Juan is much more capacious. The bottom is fine dark sand in 9 to 10 fms. Did not succeed in getting observations here which was the object of our anchoring — we therefore weighed at 11 am and steamed up the Strait of Fuca with a calm — weather thick and appearance of SEly wind — we were off the Race Rocks at 4.20 PM but just met the Ebb which being Springs ran with great Strength and it was not till 7.30 PM that we anchored in Esquimalt where we found the Topaze and the 2 GunBoats Grappler and Forward.

SEPT^{R} ESQUIMALT–VICTORIA 1860

SEPTEMBER 29, SATURDAY. Hauled 2 of our boats up to repair and made arrangements for Provisioning etc. It blew a hard gale from SE after midnight but only lasted a few hours. It was felt much more, I hear, off Columbia River and detained the packet several days.

We remained at Esquimalt during this week or until the 8th of October getting provisions, giving the Ships Company leave, and arranging about the lighting of the new light houses. The weather remarkably fine. The Steamer Brother Johnathon did not arrive until Sunday 7th, being nearly a week overdue. By her we heard of the Termagant having been placed on the Mare I^{d} Dock[80] and by some accident in the Shoring of her, she fell over and did immense damage to the Dock — being herself in a Critical position for 2 days before she could be extricated from the wreck which was done by cutting away the connexion between the pontoons. I am told that the mishap arose from error on the part of the Dock Master who takes sole charge on these occasions. It could not have been the ships weight for they have lifted vessels infinitely heavier.

OCTOBER 8, MONDAY. Left Esquimalt and steamed into Victoria harbour for the purpose of fixing on an eligible spot for a harbour light

[80] At this time, Mare Island, in San Francisco Bay, was the only dry-dock in the North Pacific.

there to enable small vessels to enter at night with the aid of Fisgard I^{d} Light. At 1 PM anchored in James bay & steadied the ship with a hawser.

OCTOBER 9, TUESDAY. I took the harb^{r} Master to a spot which I consider the best adapted to the purpose of a harb^{r} Light — and called on the Governor afterwards to acquaint him that I saw no objection in a nautical point of view to the light being placed on Colville I^{d}. Married the 1^{t} Lieut at 11 am.

> OCTOBER 4. We were all much astonished when Wm Moriarty our 1st Lieut^{t} is to be married on the 9th to Miss Read — the last occurrence anyone ever dreamt of: the old girl after flirting with all hands trying to hook some green horn at last hobbled poor Jack: well there is no accounting for taste and she may make him a good wife. They are to be spliced the same day as John Coles who marries Miss Harcus a young lady that has always lived at the Reads house. On the 9th we went round into Victoria Harbour and hoisted the Garland a second time and at 11:30 poor old Jack was taken in & done for. At 4 PM the Plumper sailed and left the victim behind with six weeks leave . . . (Gowlland 1860–1863).

OCTOBER NANAIMO 1860

OCTOBER 10, WEDNESDAY. As soon as the tide was sufficiently high we steamed out of Victoria at 11.30 am, taking the harb^{r} Master to shew him the marks and bearings for entering Victoria harb^{r} from the East^{d} both inside and outside Brotchy Ledge. Dropped him in his boat and proceeded on — passed up Haro Channel — off Stuart I^{d} passed the barque American — bound for Nanaimo — steamed between Patos & E^{d} [East] pt[81] and shaped a course up Georgia Strait for the night. At 4 am 11th was reported to me that there remained only 3 hours Coals — we should have had 10 tons onboard. This has occurred twice and I

[81] On Saturna Island.

strongly recommend that whenever 24 hours coal are reported in the Engineers register as remaining, that a personal examination should be made of the bunker and the quantity remaining reported in person by the Chief Engineer. At daylt 5.30 am found we were set 5 or 6 miles north^{d} of our reckoning — steamed in between Five Finger and Lighthouse I^{ds} and anch^{d} at Nanaimo at 7 am. Found that the 3 coal launches were engaged in Coaling a vessel, the Georgiana, at Newcastle I^{d} so we had to take the old Skaws which will scarcely float and bring our own off from the Shore loading them ourselves. The arrangements for Coaling are extremely defective here. There ought to be a dozen launches flat bottomed holding 80 or 100 Tons each always kept full and anchored outside low water, more especially a man to superintend the loading of them. I was obliged to leave 16 tons short of coal from the tinkers on the coal pit loading a Skaw so full while aground that she sank as the tide rose over her. A party of Stikine Indians arrived today, having been attacked by the Yucultas on their way from the North. 8 or 9 murdered & 5 desperately wounded. It appears the Yucultas affected to be on good terms with them, enticed them into their houses, shewed them the papers the Roman priests had left with them enjoining them to be peacable with all men — and then gently murdered them by firing indiscriminately into them.[82] 2 women were among the murdered. Our Doctors visited the wounded — the injuries had been inflicted 6 days & the Indians themselves had in some cases cut the bullets out. Nothing could be done further without running them to an infirmary — so they must take their chance.

❧ OCTOBER NANAIMO–HOWE SOUND 1860

OCTOBER 13, SATURDAY. We left at daylight and steamed for Jervis Inlet where we dropped the 1^{t} Whaler with 10 days provisions under Mr Pender to complete the S^{d} arm, & steamed for Howe Sound ourselves. Saw several whales.

---

[82] By the 1860s, the Yaculta or Euclataw, now known as the Laich-Kwil-Tach or Lekwiltok, were in complete control of the narrow passageways along the east coast of Vancouver Island at Seymour Narrows and Yaculta Rapids. They were reputed to attack and plunder those who were passing through (Duff n.d.). This prevented any north to south movement without Laich-Kwil-Tach knowledge and gave them absolute control over the region and access to Victoria.

In the afternoon saw the Alert steaming for Nanaimo. Weather very fine and clear. Steamed into Howe Sound thro the passage formed by P^{t} Gower and the large Island eastward of it which from its shape I have called Crescent Island. Struck soundings in the centre of this passage with 4 fathoms and found subsequently that a bar composed of sand and gravel extended the whole way across from the South end of Crescent I^{d} to the west shore inside Gower point. On this bar there is as little as 7 feet at low water in some places. The Shark had examined it some days previously and reported a deep channel. At low water we should have struck. We anchored on the west side of Crescent I^{d} in a Snug Cove formed by 2 Small Islands lying off the large one. Moored NW & SE, the outer anchor in 15 fms, the inner in 5. There is room for 2 vessels of this class moored, or for one vessel of any size. A beautiful clear evening.

OCTOBER 14, SUNDAY. This morning came in with rain and cloudy unsettled looking weather — and continued so during the day. I have generally remarked that one of these extremely bright clear nights are followed by a rainy day. There are several villages in our neighbourhood, but we have only seen 2 natives in a small canoe. Stakes for drying fish are very numerous, the natives are probably higher up catching salmon, but I hear that they evinced a disposition to avoid the boats here on a former occasion. When this is the case they have generally committed some act of which they are ashamed and fear punishment. Perhaps the intelligence of our remonstrance and demonstrations towards the Cocquells at Rupert has reached here. It is astonishing how soon every piece of intelligence is conveyed from one tribe to another — even tho they may be hostile towards each other.

OCTOBER 15, MONDAY. Cloudy morning but cleared by 9 PM & the remainder of the day very fine. Commenced the Survey of the Entrance — a fine starlight night, but still a grayish tint about the sky tho the stars shone brightly thro it. Barometer little below 30.00. Soon after midnight we had a thunder storm, lightning very frequent and vivid, rain in torrents, wind shifting about & blowing in gusts. No doubt a gale in the Strait.

OCTOBER 16, TUESDAY. Rain all the forenoon. After noon it held up for 3 hours and we were able to do a little surveying work. At this season it is necessary to take advantage of every fine hour, as after this time fine weather for more than a day at a time is the exception. Last year it rained throughout October and was fine the greater part of November. The first half of October this year has been remarkably fine, but the summer was not as uniformly fine and dry as the two preceeding ones. I suspect these deep inlets surrounded by high wooded hills get a greater quantity of wet than Esquimalt and the more open parts of the Strait southward. Rain continued all night with heavy overcast sky, but little wind.

OCTOBER 17, WEDNESDAY. Heavy rain ushered in the morning which is bad for our detached parties. An epidemic has gone thro the ship during the last week which has brought our sick list as high as 20 and placed 2 boats hors de combat. A kind of catarrh, which however has not lasted long. Today the sick list is reduced to 7 men — the same has been very general at Victoria among the inhabitants and occurred during the very fine dry temperate weather.

OCTOBER 18, THURSDAY. Rain all day unable to do any outdoor work. Boats left but were compelled to return.

OCTOBER 19, FRIDAY. A fine clear morning. Hills covered with snow (above two thousand feet in elevation). Temperature fell 10 degrees — and the morning frosty. Boats at work surveying the Sound. A fiery NW wind outside, but at our Anchorage quite calm.

OCTOBER 21, SATURDAY. Another fine day, but blowing strong outside and I could not pull to the west^{d} along the outer coast as I had wished. Got observations for Lat, time & TB, and afternoon went surveying. Mr Bull returned in Gig this coming from the head of the Sound.

OCTOBER 21, SUNDAY. Rain again today, but not very heavy. Sky overcast. Yesterday Mr Brown, the paymaster, went to Burrard Inlet in a canoe, intending to cross the trail to New Westminster. Mr Browning discovered another Anchorage on the SE side of Great I^{d}.

OCTOBER 24, WEDNESDAY. During the last 3 days the weather has been remarkably fine and all boats have been employed. The temperature lowers however considerably at night and the mountain tops are fast covering with snow. The winter however is evidently later than last year — for as yet no wild fowl have made their appearance from the north.

Last October early in the month they were here in great abundance. This is an unerring sign of the mildness of the season. Yesterday got letters from Col. Moody at New Westminster. Mr Brown having arrived after spending 16 hours in a canoe late on Saturday night and dispatched the Indians back to us.

OCTOBER 25, THURSDAY. A change in the weather this morning — rain and overcast with clouds. However at this time of the year we must not be too particular as to weather. Weighed at 7.30 am and Sounded all day. The water very deep. Steep Stony hills rising almost perpendicularly from the waters edge to 3000 feet and over. Their sides have been subjected to the action of very extensive fires; and the trees are completely destroyed for miles. The blackened trunks only remaining, and leaving the bare stony sides of the Mountains entirely exposed. We anchored at 6 PM in the SE harbour of Great I^d^, an excellent anchorage in 8 fms for several ships — which is a great advantage in a place where the general depth is so great. Thus there are two good anchorages in this Sound and Mr Pender has discovered an excellent harbour at the Entrance of Jervis Inlet.

The water at the head of Howe Sound is white almost like milk and water — resembling that of the Fraser and nearly fresh 5 miles from its head, a very considerable river, the Squamisch [Squamish], emptying itself into the Sound.

OCTOBER 26, FRIDAY. A very rainy, dirty morning. Weighed at 9 am and continued the Sounding of the Inlet returning to Great I^d^ harbour at 4.30 PM. A SE gale blew tonight with great violence towards morning and we let go a second anchor but no sea got up.

OCTOBER 27, SATURDAY. It blew hard all day with continued rain and the ship was unable to move. The Baro had been down to 29.54, but began to rise slightly this evening.

HOWE SOUND–FRASER RIVER, OCTOBER 1860

OCTOBER 28, SUNDAY. Raining this morning but, as I suspect the hills by which we are surrounded are the cause of it, I steamed out into the Strait — where sure enough — we found fine weather. Wind light from SE. Shaped a course for the Fraser and arrived at the sand head at noon.

The north sand h^{d} buoy having ~~shifted~~ gone caused us some trouble & we had to examine the entrance. Passed in at 2 PM the tide about half flood. Least water is abreast the last buoy but one (white), here we had 17 feet. The Channel has not at all altered nor I think the depth. The old low head buoy on south side remains where we placed it 18 months since, as well as the 3^{d} one in on the north side. The rest are gone & have been replaced by Colonial authorities. We anch^{d} off RE camp, New Westminster, at 5 PM. The City has increased considerably and presents quite an imposing appearance. The river is high for October in consequence of late heavy rains. We swung to the flood off Westminster 3 hours after the water began to rise and remain so for 3 hours. No great strength of Ebb. I am told the flood stream now reaches Langley. The bank off the Camp not quite uncovered yet.

We remained here till Thursday morning 1^{t} Nov^{r}, raining all the time — wind Easterly. Visited the town and assay office, saw the process of melting and assaying the metal. Several of the miners coming down gave good accounts of the Gold findings. Cap Grant still working at the Harrison Dam.

> OCTOBER 29, MONDAY. Lying at New Westminster. The town since our last visit here in the Summer of '59 has rapidly improved both in size & population; they have formed a Corporation & Elected a Mayor — built a Church, School houses — treasury — and residences for the Government officials — besides numerous hotels, billiard saloons, restaurants etc. and the whole town has a thriving busy aspect, everybody appears to have something to do and no Idlers knocking about — expect perhaps a few lazy sappers and miners from the Camp.
>
> The soldiers of the R. Engineers — having known our men so long, and being very good chums — gave the Ships' Company a grand Supper at their Mess Room by permission of the Captain & Col^{l} Moody. They Enjoyed themselves very much (Gowlland 1859–1864).

---

§ FRASER RIVER–NANAIMO–ESQUIMALT, NOV. 1860

NOVEMBER 1, THURSDAY. Weighed at Daylight and Steamed down the river so strong a westerly wind blowing up that we were obliged to anchor at 8.30 am. At 10 moderated a little — tried it again. I got down to Goose I^{d} when it increased very Considerably and we again Anchored. Wind moderated at night but low water obliged us to wait till morning to cross the Shoals. Stock duck very numerous. Several Snipe also.

NOVEMBER 2, FRIDAY. Weighed at 6 am passed out of the river an hour before high water — 19 feet the least water at the head. Sent Shark to ~~Nanaimo~~ Esquimalt and proceeded on for Nanaimo. Weather fine and cold. Glass high 30.30. Having coaled we anchored at Esquimalt on the 4th November. Moored the ship and got the winter work under way in the chart room on shore.

November and December passed in the usual way with one startling exception. Poor Bull died very suddenly on the morning of the 14th of November at his house at Esquimalt — a melancholy termination to the Plumpers survey. His loss in a 'public' and private point of view is a very great one to all connected with the Ship.[83] On the evening previous a son was born to me — and his name was called Vancouver.[84] All the staff employed getting the charts completed and traced. Fisgard I^{d} Light exhibited on the 19 Nov. On the 23 December the Hecate arrived and preparations were made for shifting over to her.

---

[83] According to the *British Colonist*, John Augustus Bull had dined the night before his death with Dr. Wallace at the local hospital. During the night he began to vomit and by early morning, he was dead. A post mortem attributed his death to natural causes (*British Colonist* 1859–1862: November 16, 1860).

[84] Richards kept a home for his wife and two daughters in Esquimalt at Thetis Cottage. They left for England in the spring of 1861 after the birth of his first son, Vancouver. It is not known if Mrs. Richards and the children ever travelled on the ship with Richards. Mrs. Richards was included in the photo of the *Plumper* crew in 1861. Richards makes no mention of their presence on board the ship in his private journal but, from at least the seventeenth to the nineteenth centuries, it was not uncommon for admirals, captains and warrant officers to have their wives and children on board with them (Stark 1996: 2). Because they were not paid or fed by the navy, they did not appear on the ship's muster.

On the 13th of November Mrs R. was safely delivered of a little Son and in the Morning of the 14th about 8 o'clock Peter Wallace came on board in a great state of excitement from the shore and gave us the startling and unexpected news that poor Bull (our Master) has just died at his house suddenly and from no apparent Cause. One can easily imagine thou it is [im]possible to describe the dismay, and sorrow of such an untoward event would bring in the Ship, where he was so much liked and so well-known. Only the evening before he had been working in the office with us in the best possible health and spirits possible never dreaming that he would so shortly be hurried into the presence of his Maker; but it appeared poor fellow his time had come and he was cut off in the prime of life, at the early age of 26, having a young wife of only 8 months marriage to lament his loss — a post Mortem Examination was held on his remains — and the Surgeons could come to no definite determination as to the cause of death. The Stomach appeared inflamed but nothing that would have caused anything fatal was detected — they wish to examine the head but Mr. Langford (the father in law) objected (Gowlland 1860–1863).

NOVEMBER 18, SATURDAY. The Forward Gun boat took the remains of poor Bull round to Victoria accompanied by a funeral party of all the officers of the fleet with Marines and band — for Burial. I was sorry I could not attend but being the only officer in the ship who had not been with him all the Commission was obliged to remain & do Commanding officer. The Blue Jackets dragged him up to the Church on the field office Carriage, volleys were fired over the grave and after the funeral service was over they marched on board the Gunboat and returned to Esquimalt about 3 P.M. This is the first officer the "Plumper" has lost during a commission of surveyors — and his death has cast a gloom over the Ship which will not soon blow over. We miss his place so much in the office — his table empty — poor fellow it is hoped he has gone to a better world. This untimely event will hurry Mr Langfords intended departure for England more than was originally intended. They all will start in Jany for England per Mail — leaving Miss Langford & Harry to accompany Mrs Richards home via Princess Royal in Feby next. Mrs Phillips is to be married to old Doctor Benson at Nanaimo — so the once happy Colwood is Scattered to the breeze. Poor Mr L's affairs connected with the farm were found very complicated and his books unaccountable and I'm sorry to add, through neglect on his own part in not properly attending to business (Gowlland 1860–1863).[85]

NOVEMBER 28. We are all very busy at the office working up and preparing our Summers Survey for tracing previous to transmission to the Hydrographic office for publication; we have done a great deal more work last season than any of the previous ones — the whole of Johnstone Strait from Cape Mudge to Cape Scott on a scale of two inches to the mile — making about 900 miles of triangulated coastline — harbours and plans on 6 & 4 inches. The whole of Jervis Inlet, next north from Howe Sound and Howe Sound itself on a Scale of two in to the Nautical Mile also the Islands of Texada & Lasqueti on the same scale — and coast of Vancouver Island between Cape Lazo and Mudge a distance of 20 miles on 1 inch scale — added to which the Inlet of Quatsino is partially examined and some part entirely finished making a grand total of nearly 1300 miles of Coast line and some hundred square miles of sounding for the year 1860. We find the working up of the mountains very monotonous as from Pt Chatham in Johnstone Strait to Cormorant Island the coast is surmounted by one rugged steep ugly looking range of Mountains descending almost perpendicularly into the Sea and upwards of 3 or 4000 feet high — but one has to make up his mind to be a dormant plyant machine

[85] Edward Langford was well known for leading an extravagant lifestyle. He entertained regularly, the cost of which was billed to the Hudson's Bay Company. In 1853, Langford's expenses were almost eight times that of his annual salary. This spending continued, albeit at a lesser rate, throughout his time in the colony (Ormsby 1958: 108, 130–131; Akrigg 1977: 187–189).

for 6 months and these seemingly difficulties soon disappear under the strokes of a good thick pen and backed by a great amount of Energy (Gowlland 1859–1864).

DECEMBER. We are all anxiously looking forward to the arrival of H.M.S. "Hecate" now due some days at Vancouver Island. Various bets are flying about as to her time of arrival (Gowlland 1860–1863).

DECEMBER 23, SUNDAY. At 6 PM when no one was thinking of her or looking out, a steamer was observed Coming into the Harbour. Night glasses were immediately brought into requisition and the long looked for Hecate was discovered in the Stranger Steamer. We were of course all delighted and a party immediately manned a cutter and proceeded on board to inspect our Ship. We found her as far as her internal appearance and fitments went more comfortable and roomy than we had even expected and were quite charmed at first sight with our future home for the next 4 years. . . . The same officers take "Plumper" home except Hankin who remains in "Hecate" as 2nd Lt with a view of becoming a Surveyor. . . . They [Hecate officers] would all like to remain out but can not prevail on the officers now here to go home or exchange with them. The Surveying officers of course turn over to the "Hecate" and we are doomed I supposed to another long commission in this outlandish place" (Gowlland 1860–1863).

DECEMBER 31. Tomorrow all turn over to the "Hecate" leaving the 1st Lieutenant Engineers & Warrant officers to assist taking the "Plumper" to England. . . . We all regret leaving the old ship which has become so endeared to us from our long acquaintances together; but as we are going to a larger and much more Comfortable vessel. I am afraid the remembrances of the "Plumper" will be all that will remain in our minds.

End of *Plumper* Journal (Gowlland 1859–1864).

▲ H.M.S. *Plumper* Officers with Mrs. Richards, c. 1859.
SITTING: Sub-Lieutenant E.P. Bedwell, 2nd Master Daniel Pender, Mrs. G.H. Richards, 1st Lieutenant William Moriarity.
STANDING: Dr. David Lyall, Paymaster W.H.J. Brown, Captain G.H. Richards, 2nd Lieutenant R.C. Mayne.
(ROYAL B.C. MUSEUM & ARCHIVES: B-03617)

▲ Commander John Thomas E. Gowlland.
(ROYAL B.C. MUSEUM & ARCHIVES: I-84438)

▲ Rear Admiral Richards Charles Mayne.
(PHOTO: WALERY — ROYAL B.C. MUSEUM & ARCHIVES: A-02341)

▲ The *Shark* in Detached Service, Griffin Bay, San Juan Island. (MAYNE, 1862: 37)

▲ Nanaimo Fort and Coal Works, c. 1860. (MAYNE 1862: 35)

1861
Survey Season

ESQUIMALT 1861 JAN^{Y}–MARCH

JANUARY 1, 1861. We shifted into the Hecate — all the officers of Plumper except Mr Moriarty & Mr Croker; as P Mast^{r} Engineers & W[arrant] Officers.

JANUARY 1. Turned over to the "Hecate" this morning. Captain Richards, Lt Mayne, Mr Pender & Bedwell (Masters), Dr Wood, Brown (Paymaster) and Sammy Campbell (Ass^{t} Surgeon) — Myself, Browning & Blunden — all the officers who brought the Hecate out join the Plumper to take her back to England excepting Hankin who remains out with the intention of becoming a Surveyor. The "Hecate" is a Symondite paddler of 240 horses power; side lever action built by Scott Sinclair & Co. Tubular boilers — new in 1859 — Rather an old craft built in 1839 — but by no means ugly. She stowes 280 tons of coals.

Built at Chatham — 810 tons (registered). Stowes 6 month provisions under hatches; for her present complement of 125 officers & men. Armament 2 10m guns, 86 cwt, 4 32 p[r] Canonades, 17 cwt and 2 24 p[r] Howitzers, 2 12 pounder ditto (Gowlland 1860–1863).

JANUARY 4. Sailed for Nootka Sound to look for the Forward Gunboat — whose absence was cause of uneasiness. She had gone on 9th Dec[r] to rescue the crew of a Vessel called the Florentia — damaged by a gale — anchored on 5th at Nootka — heard no tidings of Forward. Left on 6th went to Barclay Sound, where we arrived on 7th. Visited Alberni & Ucluelet. At the latter place found Florentias Crew from whom we heard that Forward left Nootka on 24 Dec[r] with her in tow — they parted. Florentia subsequently wrecked 7 miles west[d] of Ucluelet. Took the crew onb[d] and proceeded for Esquimalt. Anch[d] there on 9th at midnight.

On 11th Plumper with an interpreter sailed for Quatsino to endeavour to find traces of Gunboat. Forward returned on 14th having borne up and run round the I[d]. Had much bad w[r]. Engines out of order — short of Coals and provisions.[86]

JANUARY 18. Plumper returned. The weather in the Sheltered ports has been fine and mild — up to this time the winter certainly much drier and milder than any we have passed. Light on Race Rock exhibited on 26 Dec[r]. A vessel the Nanette wrecked there on 22[d].

JANUARY 22. Langford and his family left for England in the steamer California. 2 daughters remained with us.

---

[86] In December 1860 the Peruvian Brig *Florencia* capsized and then righted off Cape Flattery (*British Colonist* 1859–1862: December 13, 1860). Richards sent Lieutenant Robson of the *Forward* to her assistance but when the *Forward* did not return, Richards set sail in the *Hecate* to search for her (*British Colonist* 1859–1862: January 11, 1861; Richards 1860: December 13). He discovered that the *Forward* had towed the *Florencia* out of Nootka Sound and cast her off about eight miles from land after experiencing problems. Richards brought the *Florencia* crew back to Victoria on January 10, 1861, and set out again in search of the *Forward* (*British Colonist* 1859–1862: January 12, 1861). On January 14, 1861, the *Forward* finally returned to Victoria. Her tubes and boiler had given way. Due to this and the bad weather, it had taken her fifteen days to travel from Fort Rupert to Victoria (Richards 1857–1862: January 1861; Gowlland 1860–1863).

JANUARY 25. HBCo ship Princess Royal arrived from England in 120 days. Last of January and all Feby Stormy. SEly weather.

On 28 Jany Plumper Sailed for England. Cheered her out. For the last 2 months we have been decking and preparing the Shark for detached service, making a Schooner of her. I am preparing my wife & children for a voyage to England in Princess Royal, in which ship I have taken a passage for them. 10 Feb christened the Baby (Mr Woods) 11th Feby Mrs Benson[87] came from Nanaimo — arranged that Louisa Langford[88] should remain with her instead of going home. 20 Feb broke up house — took the family to Victoria ready to embark — sent traps onbd. Mary put up at Cap Reads. Heavy SE gales. 22 Feb Lady Franklin & Miss Cracroft arrived in 'Oregon'. Came up to Victoria by our boat and returned same afternoon — sailed for the Sound intending to return here — Princess Royal detained by Heavy SE weather. Saturday Princess could not get out, heavy gale. Sunday 24th 6 am Otter towed out the Princess in a Lull and she went finally. SE gale freshened and blew almost a hurricane all day. Otter returned at 5 PM never having encountered such weather. Lady Franklin returned this morning. Came onbd Hecate. Staid Church then went to Victoria. Mayne & I in Co. Early in March she went up Fraser River. Hankin in charge, and staid 10 days.

Busily employed at the chart rooms preparing work to send to England — on the ~~16th~~ 22d March having left our box of charts here to go by next

[87] Richards named Mt. Benson at Nanaimo after Alfred Robson Benson, the Hudson's Bay Company doctor there. In a letter to Washington on January 25, 1860, Richards wrote, "the natives of these southern parts rarely give names except to the immediate locality on which they are settled, which if it is large generally obtains the name of the Tribe as for instance Cowitchin valley or Saanitch District. Any good fishing place also they have frequently a name for; but I have not been able to ascertain that there is any native name for Mount Baker even, on asking the question of a Chief he told me 'Hias siah' which merely means in Chinook Jargon 'it is a long way off' — in like manner Nanaimo mountain or Mount Benson of our Chart a native if he is asked for the name says 'Wake siah' or "not far off" (Richards 1857–1862). In fact, the Snuneymuxw do have a name for Mt. Benson, *tetuxwtun*, meaning "uncovered mountain" (Snuneymuxw 2009).

[88] Edward Langford's unmarried sister, Louise, who ran the Academy for Young Ladies which Governor Douglas' daughters attended or his daughter, Louisa Ellen, who married John Josling of the Royal Navy in London in 1857.

mail we steamed out of Esquimalt for Fraser River in order to replace the Buoys at the entrance. The Forward Gunb^{t} having gone up the day before to assist, anch^{d} at 2 PM on W side of P^{t} Roberts where we found the Gun boat — not a desirable anchorage with SW winds. Visited the Western Boundary mark in company with Cap Parsons & Gosset RE. Extreme range of tide at P^{t} R 13 feet.

FEBRUARY 23. Lady Franklin arrived in the Mail Steamer on a tour through the Western Hemisphere. She has travelled all through the United States; and came here, partly to see the place, and also to see Captain Richards and his wife, the former having served with Sir Edward Belcher as Commander of the North Star in the last Expedition that went out to resume the search after Sir J.F. [John Franklin] — she is a fine cheerful, amiable old lady about 57 — full of live and Energy — very observant of everything and always wanting to know the why & where for. She took up lodgings at Victoria and was made a great deal of by all the inhabitants both in Vancouver & British Columbia; the latter place she also visited looking for a body-guard. Hankin & Miss Sophia Craekroft also accompanies her as a niece & companion (Gowlland 1860–1863).

---

§ FRASER RIVER–NANAIMO — APRIL–MARCH 1861

MARCH 14. We have at length finished all our work in the office; the tracing's being cleared up, after no end of bother and trouble, and I'm sure I have worked myself quite ill in my anxiety to get them finished before the out of door working season began; we jumped for joy to see them finally rolled up & consigned to their box preparatory to being sent off per Mail — and now Captn is all bustle & hurry to get away to our Work again (Gowlland 1860–1863).

MARCH 23. Left and steered for Fraser River entrance at 7 am — entered at 8, Forward following. Least water on shoals 18½ feet. There is 5 feet more at high springs. At 1 PM Anchored off town of N Westminster.[89]

MARCH 25, MONDAY. Gunboat took buoys and stone moorings in and proceeded with our boats down the river and returned on evening of 26th.

MARCH 28. Dined at RE mess and on Friday 29th left the river for Nanaimo where we Anchored the same afternoon. N Westminster has not much increased in size but a good deal of clearing has taken place — and on the whole is, I think, flourishing and prospering. The revenue from Customs dues collected is nearly 50,000£ a year. Lay at Nanaimo Coaling from 29 April till 4 May. 2 ships here Coaling & no Coal on hand. It took us 5 days to get 170 tons — and that had been ordered a month before.

Left on eve of 4th, passed down Haro Channel and anchd at Esquimalt 7.30 am 5th. Found here the Bacchante, flag of Sir R Maitland;[90] Topaze, Tartar — & 2 gunboats moored. Remained here till 16 April, principally employed with the Admiral — in reference to the proposal for Docks, Naval Stores, bringing fresh water down to the harbr etc, getting rates for Chronometers.

ESQUIMALT–PORT SAN JUAN–BARCLAY SOUND APRIL 1861

On 16th [April] heard of the Yale high pressure stern wheel boat having blown up on the Fraser. Several people killed or injured. Sailed at 7 PM. Carried sight of Race Lt. from our deck for 18 or 19 miles. Anchored in

---

[89] Mayne notes: "Next day we entered the Fraser and steamed up to New Westminster, without any let of hindrance. This was subject of great rejoicing to the people of Westminster, as no steamer of the *Hecate*'s size (850 tons) had before ascended the river and it showed unmistakeably that it was practicable for large vessels to do so. So delighted were the people of Westminster, indeed, that they wanted to entertain Captain Richards at a public banquet, a deputation of citizens waiting on him with that object. This, however, he steadfastly declined, representing that all he had done was his duty, and that he had come to buoy the mouth of their river, not to feast. So they contented themselves with presenting him with a complimentary address" (Mayne 1862: 224–225).

[90] Rear Admiral Thomas Maitland was commander-in-chief of the Pacific Station. The *Bacchante* was his flagship.

Port San Juan at 7 am 17th. Got equal altitudes. Strong SW wind. Topaze passed out of Straits at 12.30. 18th April Steamed out, making running survey of coast to C Beale — weather gloomy & unfavourable. Off Cape Beale at noon. Reef extends nearly a mile off. Spot the Shark. Proceeded up, anchd off Alberni at 6 PM. Fresh southerly wind blowing up. Cap[tain] Stamps mill nearly complete.[91] Remained here until 9th May, all boats left on Monday the 22d for 14 days — during this 14 days not one without rain — almost constant & generally a heavy breeze. No sun, no sight of hills, very cold and perfect winter. Boats returned at end of fortnight all more or less damaged and having done little or no work.

Mayne started to endeavour to reach Nanaimo by a direct route from here and returned in 13 days having perfectly succeeded & made a good sketch of the country.[92]

Splendid trees at Alberni for Spars. Many hundreds of acres of magnificent land — clear. Cap Stamp has 100 acres under cultivation. This is the most desirable agricultural district I have seen in the Id.[93] A large river

---

[91] Captain Edward Stamp arrived in what is now Port Alberni in April 1860 and built a sawmill that began operations in May 1861. According to Sproat, the investors had "bought all the surrounding land from the Queen of England" but "for the sake of peace" had paid the chief of Tseshaht "twenty pounds' worth of goods" (Sproat 1987: 4). The Tseshaht people did not easily concede this land and on the following day "speeches were made, faces blackened, guns and pikes got out, and barricades formed" (Sproat 1987: 4). Sproat later acknowledged that "the Indians disclaimed all knowledge of the colonial authorities at Victoria, and had sold the country to us, perhaps, under the fear of loaded cannon pointed towards the village . . ." (Sproat 1987: 7). The mill operated until 1864 and the settlement was abandoned until the 1880s.

[92] For a full account of this expedition, see Mayne 1862: 167–172.

[93] Richards elaborates in a letter to Washington: "I am more than confirmed in the opinion I then expressed, as to the great value of the timber in this district; I saw yesterday within a few hundred yards of the beach, numbers of trees, which would make single lower masts for Line-of Battle Ships or Frigates and of the finest quality. Captain Stamp of the Mercantile Marine, who is at the head of the Company and the superintending man in this Country, has large quantities of magnificent trees, cut, ready for the mill, they will be sawn into deck plank and lumber, for Australia, Chili and Peru — France Russia, Spain, Sardinia and the Netherlands have taken considerable quantities of these spars, some 340 of which have been already exported, not from here but from the neighbourhood; At this place they are decidedly superior to any I have seen on the Island. Certainly our government could save considerably by sending some of these spars to our establishments in China where no masts are to be obtained unless sent from England" (Richards 1857–1862: May 2, 1861).

empties itself into head of Alberni — and the water at the anchorage is fresh a mile from the river mouth. Anchorage good and extensive in 10 fms. But the constant southerly wind which blows up during summer causes an unpleasant riffle. There is however shelter just below the saw mill on east side quite free from this inconvenience.

After refitting the boats we left Alberni on 9th May and sounded down the Canal, having recd our English mail the night before by Cap Stamps Schooner Meg Merilles. Anchored at noon in a snug harbr oū chŭk li sit[94] [Uchucklesaht Inlet] on west side of Canal 17 miles below Alberni.

> APRIL 19, FRIDAY. Steamed up the Albernie Canal and [anchored] off the Settlement named Albernie at 6 PM. . . . A Company of Speculating Gentlemen have started a Settlement here, and Steam Sawmill with the view of supplying spurs to the world. A grant of 32,000 acres was given them at 1$ per acre by the Government to enable them to carry out their laudable intention. Some hundred mechanics have been at work constructing buildings and putting up the mill; which cuts some 20,000 feet of timber into planking daily — a large demand has been made by the Governments of France; Prussia; and Spain; and I have no doubt they will in time realize rapid and immense fortunes for themselves (Gowlland 1861–1864).

MAY 1861 BARCLAY SOUND

MAY 10. The first day of summer. All boats again left. Found excellent lime stone here which we burnt & made lime of in Kiln constructed by ourselves. Plenty of cedar — which we are cutting into plank. Got observations. This arm of the inlet runs in a westerly direction for 3 miles — width ½ a mile — anchorage in East end of it in 11 fms. An island in entrance, pass in on either side of it. On Wednesday 15 May went away

---

[94] Browning noted that the Uchucklesaht people were "trading large quantities of Winter Salmon" (Gowlland 1861–1864).

in my boat to reconnoiter the Sound which is the most extensive and broken up place it is possible to imagine, and will occupy half the season in Surveying it. Returned on Saturday 18th as did all boats. They have had a successful fortnights work. Weather fine, tho not quite settled. On Sunday 19th much rain.

MAY 21, TUESDAY. Boats left to continue the work. Went in my own to Port Effingham, an inlet running 5 or 6 miles in an NNE direction, where it nearly joins the western arm of this place, from whence it runs WNW for 3 or 4 miles. A stream running into the head of it as is usual in all these arms — the weather fine during the week — went westward thro Sechart passage and after doing some work among the islands returned on Friday 24th. Found the Meg Merilles had arrived and brought our mail the day I left, and was hourly expected on her return to Victoria.

In many parts of the Sound, particularly in the deep inlets, are remains of Native Villages — which are always easily recognized by a green spot cleared of trees where Sting Nettles grow most luxuriously — these places have certainly not been inhabited for many years — which would seem to indicate a decrease among the Natives. The tribes shift their residence according to the season. Wherever fish are plentiful in that neighbourhood is their village; when the Salmon move up the creeks or rivers they follow them and they have at least 2 places or residence during the year. The poles of the village they leave standing carrying the cedar planks and mats which form the sides and roofs with them — there are several tribes in this Sound. The Ohiots occupy the Eastern side and are rather numerous, about ________ men. The Sesharts and Toquarts occupy the Central parts, and the Ucluelets the western.

Perhaps the whole male population amounts to ______ women ______ and children.

As far as we can ascertain their habits, they are simple enough. The providing of food for their subsistence is almost the sole occupation of their lives — they are very indolent and unless actually wanting food do not care to bring any thing for barter. Bears, Raccoons, Minx, and the Seal hair and Fur are pretty numerous. Deer also, and in the interior of the

I^d Elk in troops. Salmon they bring abundance — they take in Exchange biscuit, Tobacco and Clothes generally.

Blankets are most in request and they ask absurdly exhorbitant prices for every thing — in fact, dont care to bring anything but salmon which are easily procured, unless their tobacco or clothes run short. Almost all have some article of European manufacture as a garment, or most generally, a blanket.

Few of the men have more than one wife. Tho polygamy is allowed and practiced. The Chiefs generally have 2, sometimes more when their early partners get aged which they do at 25.

The women rarely have more than 2 children, and many have none. They like intermarrying with other tribes as it heals feuds and gives them influence. In such a case the man pays for his wife to her friends[95] from 2 to 20 blankets or a Musket, according to his possessions. The women are not always faithful — the punishment is generally a beating and desertion. The young people of the tribe generally make matches of affection, but I am told the young females are not chaste before marriage. Want of chastity before marriage is I fancy not counted as sin or want of virtue. I have observed at this place that the women are kept in the background and dont come alongside the ship with the men as they do at other places on both west and east coasts. I have seen very few women here. Every tribe has a Chief whose authority is never questioned. They appear to have grades in society very much like ourselves. The Chieftanship is hereditary in the male line and the age, even if he is a child, makes no difference in the respect and deference paid by the other members. When there is no male descendant, the female is generally married to some one chosen by the tribe and he becomes chief and takes her name — as for example Macquilla at Nootka. He married the descendant of old Maquilla and now assumes the name and chieftainship.

[95] Richards probably meant "family." This is known as bride price — a practice in which the groom and/or his family compensates the bride's family for the loss of a daughter and family member.

On Sunday morning at 4 the Meg Merilles called for our letters thro the kindness of Cap Stamp — who wrote me a note — placing her at my disposal for as long as I wished to detain her.

> MAY 18. The Ohiat tribe of Indians inhabit these Islands and this the outer part of the Sound, they are a most treacherous; thieving, lot of rascals; that cannot be trusted out of sight; altho' civil enough to us being man of wars men; but poor castaway ships crews have been dreadfully maltreated by them; they are a puny Miserable race the tallest of the tribe which numbers some 2 or 300 not being more than 5ft 8½in; live principally on fish; spawn; roots; clams; cockles etc. — and in the season obtain potatoes from little gardens of their own. . . . They have a curious custom of smacking themselves into a state of insensibility with a sort of leaf found in the woods and which when dried; and crumbled up resembles green tea; they fill a large pipe with this abominable stuff; (altho' it has not a disagreeable smell); and each take 4 or 5 puffs ejecting the smoke through their noses; passing the pipe from one to the other in order of the respect of ranks: the Chief first & so on: after smoking they all become perfectly insensible for about 5 minutes; which they appear to enjoy vastly; saying it was "caqua whiskey"; "same as rum"; I tried the pipe myself frequently and find it has no effect what ever over the nerves; not nearly so much as strong ships tobacco (Gowlland 1861–1864).[96]

MAY 27, MONDAY. It rained all day and prevented the ship moving as I had intended, or the boats doing any work. A rainy day is sometimes

---

[96] The Huu-ay-aht may have been smoking kinnikinnick (*Arctostaphylos uva-ursi*), sometimes called "common bearberry." The dried leaves of kinnikinnick were smoked by many First Nations along the Northwest Coast during the post-contact period. Later, when tobacco was readily available, some extended their supply by mixing it with kinnikinnick (Pojar and MacKinnon 1994: 67). A few people have reported that kinnikinnick has a dizzying effect (Turner 2010).

acceptable as it enables us to get at indoor work which we are loth to do during fine weather tho I am of opinion time is saved by keeping the work plotted.

MAY 28, TUESDAY. Moved out of Ouchuckliset and steamed down the Centre Channel sounding — found a very heavy sea — it was doubtless blowing yesterday outside. Got a few soundings and lines to the outer breaking reefs, and then entered a harbour which we had reconnoitred before, and which I call for the present Island harbour. Entrance narrow and a rock with 10 feet in centre of entrance. Moored within in 9 fathoms — a very good sheltered place. During the week all boats employed Surveying and Sounding, sometimes 7 parties away. Found a log of yellow cypress on the beach of this bay. 20 feet long by 3 feet diameter, well seasoned, a very fine prize. Got it in. This is a glorious wood country, altho none of the harder kinds of ornamental wood are found. The Birdseye Maple I have not seen much on the exposed coasts — it likes to grow in Sheltered places or low ground. The Dogwood which is rather handsome is scarce. I have not seen it on West Coast. It easily grows more than 18 inches in diameter and is Crooked. The Yew is rather abundant and likewise prefers the East side of the I^{d}. The White Cedar is in great abundance here, and makes excellent plank, but the Yellow Cypress beats all wood — is good for all kind of house work, excellent boat plank, close and very tough.

The weather has been fine during the past week but we have found difficulty in landing on the outside rocks and I^{ds} from the swell. The entrance of the Sound is so wide and so very exposed that it is rarely without a swell — but we shall erase many of the dangers said to exist and find many clear channels in as well as the narrow Eastern one — lately supposed to be the only safe one. We have also found many good anchorages.

JUNE 2, SUNDAY. Still at Island harbour. There are a few natives in the neighbourhood who generally live on our decks during the day. They bring some fish — Salmon & Rock Cod — and no more. Game there appears none. I have not seen the true Cod yet here, but must try for some on the banks outside.

> JUNE 2, SUNDAY. Indians in the Vicinity of the Ship are named Se-sharts; like the Ohiats they are a set of rascals — and one cannot be too cautious in trafficing transactions with them as they will steal whilst looking you in the face and so adroitly; that sometimes they will Escape detection and Assume a Most injured innocent appearance that would do justice to Robson. To the N and W from Seshart are the Toquarts — all connected by birth and Marriage to each other; further away to the west-ward some 12 miles and close to the Pacific; are the most power-ful and warlike tribe in Barclay Sound; named the Uclulets; numbering 500 fighting men all well armed; but like their Na-tive brothers; a set of rascally thieves — they would be only too glad to Murder all the boats crew for sake of the utensils in her; only knowing us to be a Man of War — of which they have a wholesale dread; are afraid of the Consequence. Camped amongst them, trading oil; skins etc. is Captain Stewart formerly in charge of the fort at Nanaimo, and he althou' well acquainted from years of experience with the Manners and Customs; is sometimes obliged to go very silently at night for fear of being Murdered for the Sake of the Paltry beads, hatchets, etc. with which he trades with them (Gowlland 1861–1864).

JUNE 6, THURSDAY. Since Monday morning the weather has been very unfavourable and although all boats have been away, we have been able to do but little. Strong SE gales with almost constant rain. Low baro 29.60 to 29.70. The anchorage however is quite sheltered and no swell felt, tho a heavy sea is breaking over every rock and reef close to us.

JUNE 8, SATURDAY. All boats on board except 3^{d} Whaler.

JUNE 10, MONDAY. Unmoored early, left the Shark & Pinnace behind to work and sent orders to 3^{d} Whaler to examine the coast between Cape Beale and San Juan and to rejoin the ship at the latter place on the 24th inst. Steamed out of Island harbour at 7 am and passed out of the Sound between Entrance Rock and the Western reef. When clear proceeded along across the entrance of the Sound to the westward, and then off-shore due west to commence a line of deep sea Soundings — having gone about 40 miles west. Steamed due south until 25 miles south^{d} of Cape Flattery, and then due East until within 2 miles of the coast — where we were about 3 PM on Tuesday 11th.

> JUNE 11, TUESDAY. 11:30 Met a Large Number of Canoes filled with Halibut some 10 or 12 miles off the Shore (Gowlland 1861–1864).

JUNE 11, TUESDAY. We now commenced a series of North & South lines to fill the space thus traced and sounding every 20 minutes. Passed the Tartar bound for San Francisco this afternoon. We were occupied with these deep soundings until Tuesday evening when we proceeded to the head of the Alberni for observations preparatory to our return to Esquimalt. On Saturday 15th, Pinnace returned. From Monday 17th June until Saturday 22^{d} we remained at Alberni, painting and putting ship and boats in order. The Shark & 2^{d} Whaler with Mr Gowlland and Mr Blunden were left here to complete a few soundings and then to proceed to Clayoquot and Commence the Survey of that place, and on Monday at 4 am we started for Esquimalt, towing after us a spar for a topmast for the Bacchante — sent to the admiral by Capt Stamp.

A very fine day; carried a line of Soundings within a mile of the coast to San Juan where we picked up the Whaler with Mr Browning and then proceeded up the Strait. The summit of Cape Beale about 120 feet high appears a good site for a light for the North Side of Fuca Strait — also as a harb^{r} Light for Barclay Sound. Saw the Race tower 11 or 12 miles off and passed a little after 7 o'clock, mooring in Esquimalt harb^{r} at 8.20 PM. Bacchante & Forward gunboat here. On Tuesday morning, 25 June, the Pacific steamer arrived but brought no mail. It had been sent from San Francisco in a merchant barque 'Isle of France' and may not reach for a fort-night.

Remained at Esquimalt until Tuesday 2 July. The week was chiefly occupied by me in visiting Lt Houses or going to & from Victoria on Lt

House or other Colonial work. Attended a meeting of the Volunteer Rifle Association.

On Tuesday 2 July dined with the Adml, met Governor & officials, and at 10.30 PM steamed out of harb^{r}, having onb^{d} 42 Iron Boundary monuments and Cap Grant. Steamed up Haro Strait and at 7 am anchored at Semiahmoo Bay and landed 16 of the monuments at the parallel, the water being low they had to be carried over mud & thro water for nearly ¾ of a mile. I steamed across to P^{t} Roberts and dropped the Cutter with 3 more monuments on the parallel on its E side and returned for the party at Semiahmoo. At 3 PM had both parties onboard and proceeded on for Nanaimo where we arrived at 8.30 PM. Grappler Gunboat here.

On Wednesday morning put 23 monuments onb^{d} here for transmission to Sumas River in the Fraser. Commenced coaling — had to take the Dunsmuir pit as the lighters were already loaded. Very good coal but 7£ per ton, 1 more than Newcastle I^{d}. Grappler started for Sumas up at 4.30 am Friday, taking Cap Gosset.

NANAIMO — 1861

Comet seen on ________ remained at Nanaimo until Friday 12 July.[97] Received 200 tons of coal, completing up to 300 with a few tons on deck. The Grappler having returned on Thursday morning with Mr Browning, we were ready to leave, and on Friday at 5 am we steered for Komox and anchored off Maple P^{t} [Mapleguard Point] in 5 fathoms at 11.30 am. Went on shore to cut Maple, which employed the people till 9 PM. It was high water this evening about 8 o'clock and the Ebb ran to the South until 9^{h}. The greatest strength of the stream about 2 kts.

DISCOVERY PASSAGE, JOHNSTONE STRAIT — JULY 1861

JULY 13, SATURDAY. 6 days after New Moon. Weighed at 4 am and passed thro Lambert Channel, steering to pass 2 miles outside Cape Lazo, Sounding. Passed Cape Mudge at 10.10 am, an ebb tide running about 2 kts with us to the north^{d}. I believe the stream changes at Cape Mudge

---

97 The Great Comet of 1861, also known as C/1861 J1, was first visible to the naked eye on May 13, 1861, and remained visible for ninety days (Yeomans 2007).

at or near the moment of high & low water, but west[d] of Seymour Narrows, ebb runs 2½ hours after low, & flood the same. 11 am passed thro Seymour Narrows, a tide of 3 knots (ebb) with us, and as usual, confused rips and swirls in the Narrows. Our steam was low and we found it difficult to steer in consequence. A ship should pass thro under good power — it is better to keep just east[d] of the rips in the Narrows. Passed Knox Bay at 1.30 PM good after the flood from the north[d]. Made but with no great strength.

At 3 PM passed thro Race passage west of Helmpken I[d], saw some kelp fixed just East[d] of the I[d] which must be examined. At 6 PM abreast Port Harvey — about 7 PM the ebb made in our favour perhaps 2 kts and at 9.10 PM we anchored in Alert Bay, Cormorant I[d], which is a very excellent stopping place abreast Nimpkish River in 7 fms.

JULY 14, SUNDAY. At 4 am left for Fort Rupert where we anchored at 7 am. The Natives had carried the intelligence of our approach there at 5 in the morning. They were evidently uneasy at our appearance, probably in consequence of the recent display against the Hyders by the Gunboat,[98] and as usual when they are conscience stricken or in a fix, they go to the Company's forts for Comfort. They were anxious to know if we were going to mameloose (kill them) — and were informed that we were not while they behaved themselves.

---

[98] It was reported that a group of Haida, on their return to Haida Gwaii, had robbed some settlers at Saltspring Island and "plundered" the schooner *Laurel*. Lieutenant Charles Robson, accompanied by Magistrate Franklyn, caught up to the Haida on Vancouver Island just south of Cape Mudge where Adam Horne, a Hudson's Bay Company representative, and Constable Edwin Gough were sent ashore to ask the chiefs to return to the ship. The Haida refused and "seized hold" of the unarmed Horne (Robson 1861: May 20). Robson fired a warning shot over their heads and the Haida replied with a volley of musket fire. Robson returned fire. Shortly after, Jefferson, a Haida chief, asked for a ceasefire, reporting that four Haida had been killed. Jefferson was placed in irons and held hostage until the other chiefs surrendered. The next day, the Haida camp was searched for stolen items and a number of articles were seized, including a theodolite, silk, a desk and construction tools. Those found with stolen items were detained while the others were allowed to leave with their guns (*British Colonist* 1859–1862: May 22, 1861). According to Jefferson, the *Laurel* was robbed because the Haida had discovered that the sailors had cheated them in a trade deal and that some of the "stolen" goods came from his daughter who was living with a white man in Victoria (Jefferson 1861).

Fort Rupert is almost deserted, not above 100 people here including women & Children. They are mostly at Victoria. Some on the west coast trading for Skins, which have greatly gone up in price; the HBCo now give 40 Blankets for a Sea Otter instead of 16 as heretofore.[99] Mr Lewis who is in charge of this post is absent for a few days and the Company's establishment is decreased to 3 white men, a pretty good proof of the inoffensive character of the Indians. The pickets of the Fort are so Rotten that half a dozen men might easily knock them down. The house itself is a very good and well built one — all else is in a state of decay. Got 5 logs of yellow cypress here — each log about 18 feet long by 15 to 18 inches in diameter. This wood is evidently not scarce, but it grows some distance from the beach. It took us 2 whole days to get out these 5 logs — wrote to Mr Lewis asking him to make arrangements for getting some of this wood out by native labour — so that we may be able to bring it into general use for boat building and repairing; it is the only wood in the Id available for such purposes and is very superior. Caught some of the true Cod here with hook & line. Preserved one for the Governor.

§ 1861 FORT RUPERT–SUCHARTIE HARBR

JULY 17, WEDNESDAY. Got forenoon Sights only and at 2.30 left Rupert. Sounded westward thro Goletas channel and at 6 PM anchored in Suchartie Bay which is a good enough stopping place. Water deep, but on the eastern shore 12 fathoms may be got 1½ cables from the shore. The Dillon rock which covers at ¼ flood is dangerous to a vessel without the chart. It lies at the entrance of the bay rather on the east side. The remains of a deserted Village[100] at the head of the bay — on with the East peak of a remarkable saddle[101] (immediately over the head of the bay) leads on to the rock. The Village on with the west pk of the saddle

[99] In 1837, one sea otter pelt was worth eight blankets and by 1847, one pelt was worth twelve blankets (Mackie 1997: 286). Richards notes that by 1860, one pelt was valued at forty blankets. This inflation rate reflects the massive decline in the sea otter population.

[100] This is the village of Khatis or *dzudzadi* at the head of Shushartie Bay. In the early days of the maritime fur trade, this was the principal trading location at the north end of Vancouver Island and was known to traders as both Shushartie and Newitty. There is evidence of a village and fishery at this location (Galois 1994: 297).

clears it well to the westward. A vessel from the East^{d} may pass close round the peaked rocks off the east point of the bay — and East^{d} of Dillon Rk — we passed out that way. When the rock is uncovered there is no difficulty. Several of the Nawitti Natives visited us. Nawitti is a Village near Cape Scott,[102] eastward of it, and there is anchorage off it in fine weather.

JULY 18, THURSDAY. 2 boats Sounding the Nawitti Bar which was but partially done last year. Got afternoon sights. The sea was very smooth on the outer coast today and boats could have landed anywhere. Very difficult at the time of our last visit in Septr 1860.

1861 GOLETAS CHANNEL

JULY 19, FRIDAY morning. At 4 am weighed — signs of a fog which is very prevalent on this north coast. It increased a good deal and we dropped our anchor in 9 fms on the Nawitti Bar. A strong flood running to the East^{d} of 4 kts at least, and our anchor must have dropped on the very eastern edge of the eastern bank for it immediately dragged off into 35 fathoms. The ship soon commenced to drag, so we weighed with some difficulty and steamed westward against the tide, the fog lifting a little.

I intended to have landed on Cape Scott for observations, but we are just too late for latitude and the sun became obscured and fresh SE wind sprung up, giving no hope of time or true bearing in the afternoon. We therefore passed between C Scott and Cox I^{d}, a good wide passage, and shaped a westerly course south^{d} of the Triangles — making such an examination of them as the unfavourable nature of the weather would permit. At 5 PM being west^{d} of the outer Triangle, shaped a course for Woody P^{t} and went under easy steam, sounding at Short intervals.

[101] Richards is referring to Shushartie Saddle, a double-peaked mountain about 1900 feet high at the head of the bay where the peaks of the saddle line up with Dillon Rock at the entrance to the bay.

[102] Richards likely meant Sutil Point or Cape Sutil, which he named Cape Commeral in 1862. In 1905, the Geographic Board of Canada restored the name Sutil Point (Akrigg and Akrigg 1997: 51).

At 4 am, Saturday 20th off Woody pt a very remarkable bluff Headland with a Steep Rocky Islet lying immediately off it. SEd of this Cape the shore appears much indented but thick squally weather with rain prevented my taking so close a view of the coast as I intended. Proceeded, however, pretty close along the land, Sounding and fixing such points as we could — found our running survey agree very well on connecting it at Friendly pt, Nootka fixed by former observations on our visit there last year. Passed Nootka at 2 PM. I dont think the great bank of Soundings extends NW of Woody pt, at least for any distance off Shore. It was 8 PM before we were abreast the NW entrance into Clayoquot Sound and being nearly dark and not seeing our boats which we left here; moreover there being no similarity between the entrance as appearing to us and as shown on the old chart and numerous rocks and reefs stretching a considerable distance off — I stood offshore under easy steam adding to our Soundings.

1861 CLAYOQUOT SOUND

JULY 21, SUNDAY. At 4 am we were close in with the land again abreast the Southern entrance of Clayoquot, and firing 2 guns. Our Whale boat came out with Mr Gowlland. From him I learned that there was only 15 feet water in through that Channel at low water — which it then was — so we steamed back to the northern entrance and under his guidance entered the Sound.

The natives who boarded us at the Southern entrance before our own boat assured us there was a passage for the ship — and if I had taken their word I should inevitably have run on shore. After threading a passage good enough but with numerous reefs on either side, and passed thro one or two narrow passages and over some shoals of 4½ fathoms, we anchored in a very good berth 4 miles within the Sound under Mr Gowllands pilotage, and moored the ship. I secured some not very good forenoon sights for time, and very soon after it came on cloudy and rainy. Prepared 3 boats for leaving early tomorrow morning, two for the western or Rafael Arm, and one for the Central Northern one.

JULY 22, MONDAY. Raining heavily, boats prevented from moving — employed in the Chart Room all day. There appear to be a considerable number of Natives at this place, and quiet, well conducted people.

JULY 23, TUESDAY. Four boats went westward, two between Flores I^{d} and the main; one eastward, and one southward. Shark caulking as usual whenever we rejoin her — her deck for the sake of economy (a very false one) had been laid with plank cut by ourselves which had not time to season. I might have bought seasoned plank for a few dollars — and thereby saved days of the boats work and constant discomfort to the crew. The only piece of extravagance I have indulged in during the commission has been in paying 35£ for a whale boat — which boat has added at least one fifth to the amount of work done by the Survey. I believe more. Therefore I'm for extravagance in future — tho it is somewhat late in the day to commence. Parties on shore burning lime from shells to save the Queen a little — sawing timber & cutting shingles for marks. Pinnace with Sharks crew and some additions sounding the neighbourhood.

> JULY 23. Point Raphael [is] known by the Indian name of Manosat [Manhousat]. . . . The Indians who inhabit this part of Clayoquet Sound where the Ship is lying altho' of a different name are still all the same kindred as those that inhabit the large Village of Echachets: they are named the "Ahousats"; the Chiefs name is Wack-la — a fine athletic strapping fellow about 6 feet high; they are very civil and bring great quantities of Salmon to trade; there are three or four kinds at present being traded: the Winter Salmon which are just going out are by far the best flavoured, being of that pretty pink colour like our Scotch Salmon; they are superseded by the White Salmon; the fish when boiled is nearly the same colour as a Cod; very good but not so palatable as Winter Salmon: then there are the Hook nose; & Hump-back; which we only use when on Salt provisions and no other sort are to be obtained (Gowlland 1861–1864).

JULY 25, THURSDAY. Went to Entrance, measured base, and got the means of putting on paper the work done by the detached parties, which is considerable and will help us much in the seasons work. 2^{d} Whaler returned this evening.

JULY 27, SATURDAY Evening. 3^{d} Whaler returned. Prepared & provisioned 4 parties for 12 days detached service, and having now secured the necessary observations for time, lat etc. I propose leaving on Monday morning 29th to reconnoitre the Inlet — a series of Inlets generally.

JULY 29, MONDAY. Shark started for the outside coast to Sound, Mayne with 2^{d} Whaler & Pinnace, the latter under the Gunners Charge who is now pressed into the surveying service as a Sounder. Left for the Brazo de Tofino of the old chart; and myself — with Dr Wood as my mate also a volunteer for surveying work — left at same time for same destination. Passed Eastward thro the land as shewn in the old chart and reached the sea near Port Cox at noon. Then went northward and camped at evening on west side of Tofino arm, which is very unlike what is shewn on the old chart — afternoon had [illegible]. Mosquitos very troublesome — which is rarely the case on Vancouver I^{d}, tho British Columbia is as bad in this respect as any part of the world.

JULY 30, TUESDAY. Started at 7 am and continued the examination of the Brazo (west side) to its head which we reached before noon & got a latitude — which placed it about 17 miles south^{d} of the position as shewn in the Admlty chart — and in all respects totally unlike it. Clayoquot Sound differs in most respects from any other part of the I^{d} we have hitherto visited. All the lower or outer parts are shoal and sandy instead of deep and muddy — with numerous shoals and banks. The narrow upper arms partake of the character of the other sounds, except in their geological features. Here granite is found in situ and there are great indications of the existence of metals. All the rocks have a remarkable appearance — as if their tops had been whitewashed and the white wash had been allowed to run down in streaks. On close examination this turned out to be granite with considerable quantities of quartz. Some of the smaller I^{ds} are wholly of granite. I wish we had time to make a minute geological examination, or had a competent geologist onb^{d}, for I feel very certain that the precious metals will be discovered here. Game is very scarce — one Plover is the amount of our sport in 4 days. We returned to the Ship on Thursday eve^{ng}, the 1^{t} Augt. The natives give information of an extensive lake between this place and Barclay Sound, which is entered by a rapid river only navigable for canoes.[103] I am inclined to think from their sketch that it empties itself into or near Port Cox. They are salmon fishing there at present and I shall probably send a party to explore it.

AUGUST 4, SUNDAY. Fine day, but close. Barometer going down; at 6 PM at 29.83 and every appearance of bad weather. The tribe about a mile below us are shifting their village close to us. I suppose for their own convenience. Instead of going to and fro twice a day for the[y] spend the greater part of their time lying alongside the ship-trading berries for biscuit and whatever they caught — and as our men are fools enough to pay higher prices than the things are worth, it is worth their while to stick to us. In shifting a village, they place the large cedar boards which form the sides of their huts across 2 canoes, forming a platform of them. On these boards they stow all their boxes and household goods and so shift very easily.

We have parties on shore burning lime from shells — and observing the tides.

AUGUST 5, MONDAY. I intended to have gone away today for 3 days work but the bad weather which set in last night continued with constant rain — and prevented me. The Shark returned at noon, not having completed half her soundings owing to want of wind or fog.

AUGUST 6, TUESDAY. Went to the eastern arm and returned on the eve of the 7th. Found a mail brought by canoe from Barclay Sound, and heard of Maynes promotion to Comdr.

Remained at this anchorage until Thursday 15th, on the morning of which day we steamed down to the entrance, picked up 1t Whaler on the way and Sounded along the coast eastwardly to Barclay Sound; entered the eastern channel at 4.30 PM and proceeded up. Saw the Marcella Brigantine & Surprise Schooner anchord at Saws Bay anchorage (Entrance anchorage), laden with lumber from the saw mills outward bound.

---

103 Likely Kennedy Lake and Kennedy River, draining into Clayoquot Sound at Kennedy Cove.

Dropped 3^{d} Whaler, Mr Browning, to get a TB^{g} & examine Summit of C Beale as a Lt Ho site; anchored off the mills at 8.30 PM. Met Cap Stamp on Meg Merilles going to Victoria. Got observations on Friday and Saturday. Visited the saw mills and saw the working of the saws — 20,000 feet a day; when another saw comes from San Francisco they will do 50,000 feet.

1861 BARCLAY SOUND–1861 FUCA STRAIT

On Saturday [August 9], after PM sights steamed down the Sound, picked up 3^{d} Whaler and proceeded west^{d} sounding for the night, intending to pick up 2^{d} Whaler. Mr Gowlland off Nootka next morning; a fresh NW wind all night increased to a strong wind on Sunday morning and we made but little way. At 9 am saw whaler standing out from under p^{t} Estevan. Stood in & picked her up. A series of reefs & rocks above water extend off p^{t} Estevan and make a good shelter for boats on the beach about a mile south of the pt. Shoal water extends a mile off — we got 18 fms just over that distance.

We now shaped a course along the land for Cape Scott, sounding. Wind increasing and at noon making very little way with a heavy sea. My wish was to get round C Scott & anchor for observations there on Monday morning, but this I now felt was impossible and as the NW wind would make a sea on the Coast which w^{d} prevent boats landing for several days, I determined to alter my plans and at 2 PM bore up for Fuca St to complete some soundings there. Ran along the Coast to the East^{d} about 10 miles off with a fresh NW wind and full moonlight, fixing the ship on the chart every 2 hours. About 11 PM ran into a fog suddenly. Continued the same course along the land until 2 am when we should have been abreast the Northern Entrance and about 5 miles offshore. The speed was then slackened, and I went on deck & sounded in 53 fathoms, the sounding agreeing with the ships supposed position. She was then kept along the north side of the Strait as I believed, stopping and Sounding every half hour. At 4 am Sound in 19 fathoms — and as I knew of no depth of that kind but close to the north shore of the Strait, I judged myself to be very close to it and steered due south for 1½ miles, then hauled along the land again in 36 fathoms — sounding every 10 minutes and stopping for the purpose. Speed reduced to 5 knots. At 8.35 am

I went below to breakfast, leaving Mr Bedwell Add Master looking out. 2 men in the chains with hand leads, and same orders to stop and sound every 10 minutes. There was 48 fathoms when I went below. In 2 minutes I heard the order given "hard a port" — "Stop her", and came on deck in time to see her strike between 2 rocks just above water. Fog so thick that we could not see 20 yds.[104] Ship pounded heavily aground amidships and found calm but a swell. Lowered all boats, and got anchor & hemp cable ready. Hung anchor to a paddle & towed her astern. Dropped the anchor. Difficult task on acct of tide and swell. Tide rising. A canoe came with Master of a coasting vessel, the Elizabeth of Port Townshend, James Melvin. He told me we were between C Flattery and Neah Bay, while I had supposed myself on the north side of the Strait and some miles eastward of Port San Juan. He also informed me that the water would rise sufficiently to go within the rocks, and as the ship was forging ahead on them, I steamed ahead, slipped the cable & got into 7 fathoms, dropping a bower. There was not room to swing. Made fast with hawser to rocks. Ship making water about 6 inches in an hour having struck heavily on both bilges under paddle boxes. Put boats on the sunken rocks (with the help of Mr Melvin) between which we were obliged to pass in getting out and after some difficulty and considerable risk threaded our way out. At 11 am up boats and steamed for Esquimalt, the bilge pumps keeping her free. A few minutes after we got off we ran completely out of the bank of fog — into clear weather. This fog had hung about the Entrance of the Strait for 5 or 6 days, as I was informed by Mr Melvin. The immediate cause of the ships grounding was keeping her South of 1½ miles after the 19 fathoms were struck. A few minutes after the 19 we struck 28, so that I felt no kind of doubt but that we were on the north side of the Strait and very close in. Is it a bank

---

[104] Mayne describes the *Hecate* crew's shock and surprise at running aground: "Nothing but rocks were to be seen all around us, and we were all equally puzzled to know where we were, how we got there, and how we should get the ship off. That we were close to the shore we soon found, for high up over our foretopmasthead, as it appeared from aft, the summit of a cliff, with a few pine-trees upon it, showed itself. Fortunately for us, the noise of the steam escaping was heard by the master of a small schooner, which we afterwards found was lying close to us, and we soon saw two white men . . . paddling to us in a small canoe. Getting them on board, we discovered that we were two miles inside Cape Flattery . . ." (Mayne 1862: 237).

in the middle of the Strait of which there is a rumour from Indian report. A strong southerly set must have existed beyond any ordinary or spring tide — and I expect that these strong sets must occasionally exist after certain winds. At the same time of the moon in Sep 1860, the Plumper was within a mile of the same spot in what our reckoning placed us at 4 am of 19th Augt and experienced but little tide any way.

Next Morning Monday the 20th August 1861 — My Morning watch and one of the densest fogs prevailing I ever remember seeing. By our reconning we ought to have been close in to the South Shore of Vancouver Island. Went on very slowly all the morning in consequence and Sounding repeatedly, that is every 10 minutes; at one bell I was relieved by Bedwell our Additional Master and went down to Enjoy the nice little breakfast, the Midshipmen of the Watch (young Sulivan) had been begging and stealing all the morning. We were just sitting down to it when a bump, bump and Crash, warned us that breakfast would be a myth; up we all rushed and there was the poor old Ship in a rock amidships, and rocks all around her close to the gangway. Every rise of the sea which was a heavy swell only drove her further on, and the heavy shock as she settled again on the rock nearly threw us all off our legs & threatened momently to send the Topmasts down about our ears — the boats were immediately all lowered and paddle box boats got out and an anchor immediately got out astern, which of course took some time perhaps — 10 or 15 minutes, the old Ship Thumping heavily all the time. Every instant we expected to see her break up amidships; and no doubt she would have had not a Yankee Captain from a Trading Schooner lying close in to us, put off and told us that over the rock once safely we should be in Deep water and safe — the bunkers by this time by the Concussion had all come in and the side lever of the Engine brought up in its stroke By the large Combing of the Hatchway but thro the prompt energy of the Engineers it was immediately cut away and the Engines our only Chance moved ahead; and

> to our joy and relief when we had imagined all lost she went over the rock into deep water. Slipped the Cable and steamed safely in Esquimalt that Evening making about 8 inches of water an hour — found Bacchante at anchor (Gowlland 1860–1863).

1861 ESQUIMALT

The "Sierra Nevada" arrived at Esquimalt on Monday 26th Aug^{t}, just a week from the day we got ashore, consequently at the Neap tides & we believed she had passed C Flattery light 6 miles and hauled into the Strait East^{d}. The fog was still thick. Suddenly the fog lifted for a moment and the land as near as possible — the same seen & run on by us was observed ahead & a few yards off so near that when stopped she was within a stones throw of it — and had a most narrow escape, of ship and all hands.

Archdeacon Wright who came up passenger with "Sierra Nevada" described the occurrence to me. The entrance of Fuca Strait is most dangerous to make in foggy w^{r} and during Augt Sep^{t} Oct & Novr such fogs as we had are common, sometimes hanging in one spot for a week if it should be calm — but especially is it dangerous coming from the north^{d}, as the Ebb besides running out of the Strait sweeps along the Coast to the South at an uncertain rate, varying according to the strength and direction of gales which have been blowing. As I believe these gales are the causes which produce the uncertain currents [they] may be comparatively remote from the mouth of the Strait. For instance, the day before we ran ashore we had a NW gale we could not steam against abreast Nootka Sound, while at the Entrance of the Strait not 100 miles dist there had been no wind and a fog for 5 or 6 days.

1861 ESQUIMALT TO SAN FRANCISCO

From the 19 to 31 August we remained at Esquimalt — fire lit under one boiler to keep the Donkey Engine going which kept the ship free. The diver named Kelly from Bacchante with a Diving dress commenced to patch the bottom up by stuffing tarred oakum grease and felt into the fractures. Over this was nailed blankets, and finally sheet lead. The leaks gradually subsided as he completed his work and by the 30th were reduced to ½ an inch per hour. Mayne went down & examined all the bunged parts & considered that they were very skillfully stopped. The starboard side appears to be the most serious injury & there is reason to think some of the floors & frame work is shattered & the bottom of the Coal bunker is lifted up — this we shored down from upper deck beams in 2 places with long spars — and wedged up tightly. We filled with coal and started on Saturday morning the 31^{t} Augt the Mutine in Co — as convoy. Had a calm run out of S^{t}. Baro. 30.16 — water smooth; rounded C Flattery at 3 PM and getting an offing of 10 miles steered SE Mag. Fine night, a slight rolling swell. Sunday fine w^{r}. SSwesterly wind, and a little swell. One of the long Shores worked out this morning, the Starb^{d} side. Stopped & wedged it up again, then went on. Fine all day. No current since leaving Cape Flattery.

SEPTEMBER 2, MONDAY. Very fine calm morning, sea very smooth. At 9 am signaled "Mutiné" to part C^{o} and join the Admiral. Capt Graham came onb^{d}. He informed me he had found the coal taken in at Esquimalt very inferior and had been burning 20 tons daily instead of 12 — so that had I kept him longer, he would have had to go on to San Francisco to coal, which would have entailed great expense — as well as inconvenience — risk of crew leaving etc, and having no apprehension whatever for the ships safety. I had no hesitation in directing him to leave us. By our noon observations we had felt no current. The land about Cape Blanco seen bearing E½S distant about 50 miles, so that the land must have a considerable elevation. Fine weather and very light SEly wind up to 10 PM and continued during the night.

SEPTEMBER 3, TUESDAY. Morning gloomy but little wind and that SEly. Saw the land on the port side and Cape Mendocino ahead — found by lat and bearing of Cape we were set 8 miles to Eastward and as we were still being set into the bay North of Cape Mendocino, we kept out 2 points. The land about the Cape high and bare of trees, generally smooth and bearing the brown burnt appearance so characteristic of Californian scenery and in such strong contrast to the pine clad region we have just left.

Cape Mendocino is remarkable — a projecting but rounding headland. Off its pitch is a remarkable small roundish looking I^{d} of the same colour as the main, but much striped with white.[105] Several white cliffs, or perhaps yellow would be better — occur just north^{d} of the Cape itself. The land 3 or 4 miles within rises between 3000 & 3500 feet. Immediately off the I^{d} just mentioned lies Blunt Reef 4 or 5 miles offshore. There is a passage between it and the main, which the American mail steamers have been accustomed to take — many of them have run great risks by so doing — and 2 years ago the "Northerner", a vessel of 1500 tons commanded by Cap Dall, the best seaman & Pilot on the coast, was totally wrecked and many people perished — she struck on a rock which was known to uncover at lw — but so great is the desire to cut off corners by these dollar making people, that they run all risks — nor think of passengers or themselves. At 3 PM abreast C Mendocino & at 3.30 steered SEly to pass 20 miles off P^{t} Arena. Weather very fine up to 10 PM, glass steady. The breeze all day has been rather a fresh sea wind from about South. A dense fog, the curse of this coast, enveloped us at 11 PM and lasted more or less thick till 6 am.

SEPTEMBER 4, WEDNESDAY. When it cleared and we saw the land on our port side distant 11 or 12 miles — wooded and pretty high — the lower ranges clear & grassy — P^{t} Arena is rather remarkable at this distance, being a bare grassy piece of land falling abruptly to the water in steep yellow clay cliffs from 100 to 200 feet high, lower at its northern end. We found by bearings that we had been set somewhat inshore as yesterday.

Got observations at noon which placed us 34 miles from Punta de los Reyes. A light breeze on Starb^{d} side to which we made all sail but immediately after noon the fog again came on very dense, although the sun was out above. Blew the Steam Whistle constantly to warn any vessel of our position. These fogs are most dangerous to a stranger. To lay to would be useless, for they are such constant occurrence that one would never reach his port. The coast also is very scantily sounded. Indeed there may be said to be no soundings which will assist a vessel in determining her position. Today and yesterday we have been 7 or 8 miles northw as well as Eastward of our reckoning. Ran on all the afternoon in a dense fog. At 3 we had 57 fms, which decreased to 40 at 5 PM. What appeared

[105] Sugar Loaf, a 326-foot sea stack.

to be the summit of P^t^ Reyes was seen from aloft about 4 PM — but it was considerably outside where we considered it to be and at 5 PM we dropped a kedge anchor in 39 fathoms, believing our position to be 6 miles from the point. It is desirable in these fogs to have a man aloft — the land is often seen over the bank while it is invisible from the deck. A light westerly wind set in and we dragged slowly but as there was plenty of room I did not move but kept steam ready. The evening cleared overhead bright starlight, very dense below, but the fog began to condense and drop in large rain drops, which I consider a sign that it will soon disperse, & I was not wrong, for it lifted Suddenly at 9 PM and we saw P^{t} Reyes clearly 3 or 4 miles distant. Weighed immediately, rounded it at 2 miles and steered E^{ly} from Starb^{d} for the position of the Bell boat (now removed). At 1 am we were within a mile of it, sounded in 19 fathoms & steamed on the mile when we altered C^{o} NEly for entrance of harbour. Bonita P^{t} light 7 or 8 miles dist was not seen. Therefore I concluded the land was covered with fog — as proved to be the case. The stars shone so brightly down to the horizon nearly that the bright ones were mistaken for the light several times by the lookout man. Ran on 4 mi & lay for 7 miles away, 16 and 9 fms. 3 or 4 casts of 7¼ — it was high water spring tides — we must have crossed the bar undoubtedly, tho we felt no swell and had more water than shewn on the charts. The only guide under our circumstances is the bottom, which when inside the bar has Red specks mixed with the sand — and we found it so, having run the 7 miles which should place me very near the mouth of the harb^{r}. I anch^{d} in 10 fms gravel & red spots. From the Ebb running 4½ kts from the NE, I conclude we are very near and right off the entrance. We anchored at 2.50 am the day of new moon. The Ebb ran till 7.15 am of 5th, and after ½ an hour Slack water she swung to flood, which was not nearly so Strong. It is LW at mid^{t} at Springs, so that there is 7 hours Ebb & 5 flood at the Entrance.

∮ SAN FRANCISCO

SEPTEMBER 5, THURSDAY. Fog very thick all the forenoon. Sounded fog whistle constantly. At 11.30 am it lifted for a few moments saw a very remarkable white rock just on the bearing Bonita pt should be. Weighed and steamed in towards it with the last of the flood. Presently got a view of Bonita Lt House which is on a high steep bluff ¼ mile inside the White Rk and is consequently much more liable to be obscured by fog than the Rk w^{d} be — as we entered the harb^{r} it cleared. Passed the Fort P^{t} and anchored. At 2 PM close off the Pacific Mail Co Steam wharf at the upper end of the Town — which is the best place for a ship of war — out of the way of Merchant vessels which not unfrequently cruize about the harbour with sails loose and anchor dragging. The port regulations amount to nothing and as to places for landing, they are not known. I found no kind of steps except at the PMCo wharf. If there are any, they are constantly blocked by ships, every one builds his wharf any shape or size he pleases. There are no laws. Buoys placed on rocks or shoals are used for weighing purposes by Merchant ships and the authorities can say nothing. All the lower part of the City is built on piles and is planked over, or was once. It is no ones business to repair it and on all occasions I have seen it there are holes big enough for horses & carts to go thro. Frequently people fall thro and are drowned. A stick and a lanthorn are essentially necessary for any one coming down to a boat after dark, and a good lookout by day time. I saw the Consul and rec^{d} a letter from the Commandant of Mare I^{d} Yard complying with the Adm^{ly} request to dock us but assuring me at the same time that the Sections were under repair and that the USS Saranac was waiting to go on. I gathered from this letter that it w^{d} be perhaps 3 weeks before I was taken on — having many things to do at San Francisco — instruments to get repaired etc, I remained until Friday 13th before going to Mare I^{d}.

On Saturday 7th [Sept] I went with the Consul Mr Booker in a Gig to San Mateo, a country village 23 miles from the City, my object being to see Lady Franklin who was staying there out of the heat and dust and other disagreeables of the City. A cold fog was blown down the Valleys on us as we drove out and the drive to me was very uncomfortable & gave me a sore throat. M^{r} B seemed to enjoy it — but this dry windy foggy season of the year is most disagreeable to a stranger. At San Mateo there are several gentlemens houses — or Ranches — as the Estates are called here with old Mexican dialect — Mr Parrot & Dr Poett the 2 principal ones. There is a fair Inn which is a Sundays resort for every one but the few who remain in the City and go to church, comparatively very few. Very fine fruit grows at San Mateo peaches, apples, pears & grapes, the latter in Green House very fine. Apples & pears very large, but not like ours for flavour nor the peach either by a long way. Remained here until Sunday afternoon when we drove in to San Francisco. I lost 4 men by desertion here — who had been seduced by a rascal who came onb^d^ on our arrival ostensibly to see an old acquaintance & was foolishly suffered to come onb^d^, but I recovered all these men thro the instrumentality of a Mr Ainsbro a member of the Police force who captured them at San Mateo. He charges pretty high — they cost 35 dollars each, but I should have gladly paid double. M^r^ A gets a warrant (blank) I presume from a convenient J.P. and inserts the names himself. He apprehends them, brings them onb^d^ promptly and no questions are asked. I suppose the J.P. shares in the profit, but any how we are the gainers by the transaction and should not quarrel as to the means adopted.

The excitement of an Election for candidates for Governor of the State had largely subsided on our arrival, a Unionist had been returned but the Secession party was very strong and if it had triumphed there would no doubt have been civil war in San Francisco as well as in the East.[106] I found I think all the more respectable part of the community — had southern sentiments at any rate were opposed in their hearts to the tyrannical mobocracy which governs the UStates — very many of the first people in position here did not hesitate to express their opinions to me that a monarchy was far preferable than the despotic sway of the majority to which every one is subject.

> SEPTEMBER 10. Four Men deserted from the Ship during the night; but were brought on board by the Police in the evening, having picked them up some 15 miles on the Road to San Mateo. The Constabulary of San Francisco are a fine, determined class of men, (a great Many of whom are Englishmen);

[106] John Downey was the California governor, a Democrat and Unionist. Abraham Lincoln, a Republican, had been elected as president in November 1860 and took office in March 1861. In response to Lincoln's election, the Secession Party, made up of pro-Confederate sympathizers predominately from Southern California, called for the creation of an independent Pacific state. After Downey won the election, the Secession Party did not succeed to power in California.

they wear no uniform so that it is impossible to know when or where you may meet them. At any great National Meetings they are always to be found amongst the noiseyest & most patriotic. Even at private parties they Manage to introduce themselves never committing a breach of etiquette; and passing for Gentlemen & Guests. We experienced great Hospitality from some of the resident families, but the general feeling of all Americans at this present time is very bitter against England, doubtless on account of the war at present raging in the States & Mr Russels letter in the times, are not calculated to soothe their ruffled tempers. Passing through the streets in uniform in broad daylight; we hear the rowdies sitting on the doors of pot Houses etc. Making Remarks about those god damned Britishers — and how they are going to "give us fits" after they have finished with Jeff Davis (Gowlland 1861–1864).[107]

⚓ 1861 MARE I^D DOCK YARD

On Friday 13 I steamed up to Mare I^d and anch^d abreast the Dockyard. At 3 PM I found the Sections under repair & no chance of my getting on for 14 days. Capt Gardner the Commandant is a naval Capt and a very obliging person. Cmd^r Green is 2^d — Cmd^r Lanman superintends the Ordnance Dep^t; Comd^r Stanly commd^r Depot — Ship Independence has a New Orleans wife very agreeable & very secession — all society is split up into factions by the disturbances in the East. Every man suspects another. The workmen in the Yard equal in social position to the officers report their suspicions of the former to the President; a Mason has gone

[107] Jefferson Davis was the leader of the Confederacy. Lord Russell, the British Foreign Secretary, stressed England's neutrality. "Mr. Russels letter" likely refers to a letter published in the *New York Times* on June 18, 1861, from Russell to the "Lords Commissioners of the Admiralty" which stated: "Her Majesty's Government are, as you are aware, desirous of observing the strictest neutrality in the contest which appears to be imminent between the United States and the so styled Confederate States of North America; and with the view more effectually to carry out this principle, they propose to interdict the armed vessels, and also the privateers of both parties from carrying prizes made by them into ports, harbors, roadsteads, or waters of the United Kingdom or any of her Majesty's colonies or possessions abroad" (*New York Times* 1861).

to Washington to bring charges against the Commandant because he prohibited said Mason from making a political stump-speech on the occasion of swearing allegiance to the State. The result of this [illegible] already is that the Commandant has been ordered to send his prinl. Sec. and his son out of the yard because the Mason says they talk secession — all decent people consider their country gone.

We went alongside the wharf, got our guns & coal out. Commenced repairs of our paddle gear which had suffered and made all preparations for docking.

On Thursday [Sept.] 19th accompanied by Dr Wood I went to Sacramento City. We rode from Vallejo, a village close to Mare Id by the stage to Bencia [Benicia] and this larger village situated on the rt Bank of the river 7 miles above the Creek on which the Dkyd is situated. There are opposition stages here as there are oppositions everywhere in this country — so we rode 7 miles with 4 horses for ½ a dollar, where as another hour we should have paid 2 dollars for a ride astern of 2 horses.

Reached Bencia at 5 PM and as we had an hour to wait for the steamer, strolled about. A large straggling village with a convent school and some good houses — the inmates of which seemed to be or might be all dead for all you see of them. Dirty like all California towns — and if it were not for the fine climate & strong winds which keep the air sweet, no doubt plague would find it a congenial spot. It is no ones business to keep streets clean. I think California is far dirtier and more meddlesome under the Americans than it wd be under the Spaniards or Mexicans. The birds & dogs are the only scavengers. There are 2 inns — "American and Solano" — vile as all American inns are kept by independent drunken extortioners — or a German named Winemann who seemed 10 years more depraved in drink — tho he was on a farm not 18 months since.

At 6 PM we embarked onto the Chrysopolis River steamer, very fine vessel, low pressure, 260 feet long, fitted with a range of Deck Saloons, something like the Christal palace[108] — on top of all a box for the Capt

[108] The Crystal Palace was a glass and iron building with a large rounded atrium built in Hyde Park, London, for the Great Exhibition of 1851.

& Steersman. Funnels are rigged out on each paddle box a convenient plan, as the smoke goes clear of the passengers. We had between 600 and 700 people onb[d]. The fare to Sacramento 70 miles distant is 3 dollars — on alternate days only ½ a dollar, by the opposition which runs at a loss to try & put the other off the line. Vessel clears 7[f] 6[in] and grounds in 2 or 3 places at low water. The river has been wonderfully filled up by the mining operations and will before long not be navigable unless dredging is resorted to. Now it is like navigating in pea soup instead of water. These go ahead people dont care — sufficient for the day is the evil — let this Company make their fortune. When steamers cant run any longer, then those that come after must find the remedy. The river is very narrow & thick fogs prevail. They never check their speed, the Cap[t] navigates by his watch and the number of revolutions and takes his chance in the darkest night or thickest fog.

The Antelope and Sacramento are the opposition boats, neither as fine as Chrysopolis. Cap Chadwick commands former, a most obliging and respectable person. The opposition boats sometimes resort to the most peculiar means of annoying the enemy. If a slow boat is ahead and a faster one coming up, on goes the Bacon or any other inflammable substance onb[d]. If this wont do, when the fast one comes near, the slow one places himself across the narrow stream by a movement of the helm. The stern boat has to stop or run into him which he sometimes does. This occurred while I was at San Francisco. A law suit followed the collision. The lawyers got work — probably one was at the bottom of the business — the Cap[ts] of the opposing boats were [illegible] by the Trial — in steps another fellow all ready and makes his market. A boat or two blows up, a few hundred people maimed or killed, "a small thing" forgotten in a week.

To resume, we reached Sacramento at 2 am Friday, had no cabins in the steamer from the crowded state of her, so sought a lodging on shore. But the fair had filled the City and at no Hotel could we find accommodation — although there are several very fine large ones there, the principal the Orleans & St. George. Remained unwashed and unshaven until 4 pm when we got a miserable dirty dig hole in the Orleans hotel. In the meantime drove in an omnibus to the fair, where we saw some very inferior horses — bulls — cows — and other animals, some hundreds of very plain vulgar women & rowdy men took several drinks perforce, having been introduced to the authorities. "Cook Tarts" — "General Jackson" etc. Returned at 4 PM covered with dust and having been jammed in an omnibus big enough for 12 but carrying 20 of a side. All the omnibus drivers are Irishmen Americanized. Curious scene occurred a short time since the drivers all carry revolvers. One fellow w[d] not move on when he had his vessel full, which gave offense to another skipper who was waiting for a cargo. Some polite interchanges of words ensued, after which one fellow fired his revolver at the other from the top of his box — several shots were exchanged when one dismounted closed & stabbed the other with his bowie knife. Killed him on the spot and then made 16 more holes in him. I met the lawyer in the evening who was to defend the murderer. He was describing the case and nature of the wounds graphically to some ladies who were cooling themselves on the verandah of the hotel (who seemed to be highly gratified by his vivid description); he guessed he would get his client thro. Called on the State Governor who resides here, a D[r] Downie, an Irishman — appears sensible & intelligent — is supposed to have Southern sympathies. Married a native Mexican very pretty and agreeable & ladylike, a great contrast to the ladies I met in Sacremento. Went with them to the Pavillions in the evening, which is a kind of exhibition of art — on a small scale. Saw some good specimens of Yankee buggies very light, neat and serviceable. Some mining gear. Very large Beetroots, Water mellons and other vegetables. Some had paintings and workmanship in Silver of a most execrable character. Floor covered with Spits which must have ruined the dress of any lady who had the temerity to walk over it — and there were many. Sacramento is a handsome city, streets wide — laid out at right angles. Some handsome trees — Accacia — & a kind of poplar which gives them a cool appearance. Many brick houses, some handsome — many very mean wooden ones among them. Fires are frequent and the law provides now that when a house is burnt down, no wooden one is to be erected in its stead. So a good many fires occur in consequence. Population about 16,000 a thriving city supplies all the mining districts with the necessaries and luxuries of life.

On Saturday [September] 21^{t} at 2 PM left in the Antelope, one of the opposition steamers. Smaller than Chrysopolis & crammed so that there was only standing room. 700 or 800 people onb^{d}. Oppressively hot, and could not have stood it but for the courtesy of the Capt who gave a few of us his own cabin on the summit of the glass house. After grounding once or twice and getting off by using the 700 passengers as shifting ballast and steaming thro one or 2 feet of mud, we reached Benicia by 9 PM where after some difficulty we succeeded in forcing our way out of her, Portmanteau in hand. Here we got into the stage again and for a dollar reached Vallejo, but after suffering severe tortures, the conveyance had double its number inside and was Crowded outside. I had a drunken man on each side of me who went to sleep with a head on either shoulder. A rather well looking young woman opposite with very thick boots on, she composed herself to sleep with her head on her sweethearts shoulder and her boots on my lap. I was pretty comfortably moored. We were more than 2 hours going 7 miles and if we had not been rescued by another stage with 4 horses halfway, should have been on the road at this time. Glad I was to get onb^{d} the ship and get some supper, for I can scarcely say I had eaten since leaving here. The hotels are so dirty and so abominable in every respect that one need have a strong stomach to sit down to a meal. Not until the 4th of October noon did we get on the Dock. The Saranac US paddle St^{r} of 1500 Tons Cap Ritchie having been taken up previously. When out of water our bottom presented a most ragged and woe begone appearance. 6 or 7 timbers broken starb^{d} side; 10 or 12 outside planks experienced shifting forefoot & part of Stern gone. Several places portside gone, eight thro to timbers but none of latter broken. Her whole port side was jagged and bruised fore and aft. She had had no false keel and her main keel was not more than 4 inches in depth. It had been patched a good deal formerly, and was now much jagged and injured — about 600 sheets of copper would require replacing. A large gang of hands were placed on her to open her out and Mr Simmonds, the Constructor, worked with much zeal and energy to forward the work. Her forefoot was replaced with Laurel 4 timbers of same and placed alongside the broken ones and bolted laterally — the necessary planks taken out and replaced by Nyon [Pinyon] pine — no end of graving pieces put in — pieces cut into main keel and a false keel of 4½ inches put on for the whole length. Additional — her bottom and upper deck caulked thoroughly, and in fact the ship was to all intents made as good as ever by Wednesday morning the 16th October when she was floated again. The whole cost of repairs & Materials 6080 dollars — or about 1220£ — out of this labour was nearly 1000£. The people being paid from 5 to 6½ dollars a day. Caulkers wages are 6 dollars. Every facility was given me by Cap Gardner, to whom I felt much indebted for his unvarying attention and civility. The Dockyard is not progressing very fast owing to want of funds, and the unsettled state of the governments. The Factory however is advancing and they hope soon to be able to cast on a large scale and do any work that may be required by steamers. The casting of our metal cogged paddle wheel segments was done here tho with some difficulty. The Bishopps Durich is at last going into its place. Six of the eight sections of the dock are now in good repair — and the other two will soon be so. When the 8 are in good order, they can take a keel of 250 feet. We were raised with 6 sections, our keel resting on 5 only. The Saranac was raised with the same number, also the Wyoming a Screw Corvette of 1200 tons. Immediately on coming off the Dock on Wednesday, we commenced Caulking upper deck which required it very badly. Got coals, guns & weights in at the same time, and on Saturday morning the 19th Oct we steamed down to San Francisco, the Saranac in company where we docked at 3 PM.

There we took in 60 Tons of Nanaimo Coal at 18 dollars a ton delivered onboard — the same we pay 6 dollars for at Nanaimo No. 3 pit. 16 tons of water at 2 dollars a ton, a little provisions. Sold our old copper about 1800 lbs at 14 cents a lb, and were ready for sea on the 24th. As the Panama steamer was due however on the 25th, I waited for the mail and a Lieut I expected by her to supersede Comdr Mayne, who left for England on the 22d. The steamer arrived on Saturday morning the 26th Oct. Mr Hand came by her on Sepr 8th and on Sunday morning 27th we steamed out of the harbr with a very thick fog, cold and unpleasant. Felt our way out and when outside the fog cleared up and we steered for Pt Reyes. A fresh NW wind springing up at 11.30 we came to Pt Reyes. Wind increased to a gale from NW and we made no way at all so at 2 PM I hove up and at 4 anchored in 6½ fms in Drake Bay. The baro was up to 30.25 during this breeze. This is a very eligible spot with NW winds. In rounding the eastern part of the P^{t} some rocks & shoal ground stands off ½ a mile — gave it a berth of a mile and anchor with the *X*mc [cross bearing magnetic course] of the pt bearing south about a mile distant. Strong wind all night.

SAN FRANCISCO TO ESQUIMALT

OCTOBER 28, MONDAY. The wind moderated at 7 am and at 9 we weighed and proceeded on our way. Found a fresh NW wind outside & a swell which impeded our progress and caused us to roll about considerably — breeze died at night and in middle watch came up from south^{d} with a falling glass.

OCTOBER 29, TUESDAY. Wind veered all day between SE and SW, misty and rainy. Baro 29. Got under sail, steam on 5th grade & throttled going 7 kts. At 8 PM the wind veered suddenly to NW. Took us aback furled everything opened the throttle & burned 16 cw^{t} the hour, we had been burning 13 cw^{t}. Soon after it veered to West and we set for and aft sails. With this change of wind a heavy westerly swell seemed to get up suddenly. We had just rounded C Mendocino which is the turning point where a change is generally met with of some kind, it has probably been blowing heavily all along the coast. We got the end of the gale off P^{t} Reyes and met the end of it again here off Mendocino.

OCTOBER 30, WEDNESDAY. Very little wind and a heavy rolling westerly swell all night. This morning it freshened again from SSE and by noon we were under all sail and burning 16 cwt the hour going 7 & 7½ knots on 5th grade. Barometer 30.10. At 8 PM Baro, 30.00. Wind freshened from SSE during night.

OCTOBER 31, THURSDAY. Running under square sails forward and boom mainsail with peak chopped (it should have been down), squally with rain. Going 9 kt, wind shifting from one q^{r}[¼] to the other. Boom jibed carried away iron stanchions and nearly lost a boat. In 2^{d} reefs [illegible] boom; Mainsail split. Set topsail — in 2^{d} Reefs of F top^{l}. Wind moderated at 10 am. Made all sail. Wind gradually hauled to SW — west and at 4 PM north^{d} of west. Close hauled. Wind steady all night at W^{d}N.

NOVEMBER 1, FRIDAY. 6 am saw Cape Flattery, the L^t just visible from paddle box 20 miles off. 8.30 passed L^t Hs and steered into the Strait. New moon today, consequently hw at C. Flattery at noon carried the flood up to Race Rock which we rounded at 2 PM — about slack water by stream but tide fallen 3 feet by the rocks. Shewing flood stream was 2 hours after HW. Anch^d and moored at Esquimalt at 4 PM. Topaze here and 2 Gunboats.

Lt Robson, Comd^r Forward Gunboat, died on Tuesday morning 5th Novr from a fall from a horse, having dislocated his spine.[109] We remained at Esquimalt until Monday morning the 11th, provisioning & giving the crew leave, the weather during our stay unfavourable & windy. On the 3^d a strong SW gale with much rain. On Sunday 10th Baro fell to 28.60. Heavy rain all day. Sunday night 29.40, but no appearance of a gale. It rained very hard and the temperature fell during the night to 38° — snow fell on the near hills. I have rarely known the glass to fall to 29.40 here without a heavy gale.

POINT ROBERTS–NANAIMO

NOVEMBER 11, MONDAY. At 9 am we went round and anchored off Victoria harbour for the purpose of getting onb^d a lot of preserved provisions and pickles etc. sent out from England for us, & also to Embark a quantity of lumber and gear for erecting a Boundary Monument on the Western face of P^t Roberts.[110] During this day and the next the weather was beautifully fine with NE wind — an invariably fine wind. The glass had risen gradually since Monday morning. Temperature at night 33°.

---

[109] On October 27, 1861, while riding in Esquimalt, Robson was thrown from his horse after a sheep ran between his horse's legs. Robson was paralyzed and died as a result of his injuries (*British Colonist* 1859–1862: October 27, 1861).

[110] Richards had requested and received permission from Admiral Washington to sail the *Hecate* to the Sandwich Islands (Hawaii) for the winter of 1861–1862. On November 4, 1861, he wrote that he had "decided not to take advantage of your permission to winter at the Sandwich Isles this year but to remain here or at some other port in Vancouver Island where the interests of the survey and the Colony can be best served" (Richards 1857–1862). In November he and his crew assisted with the boundary marker construction at Point Roberts.

NOVEMBER 13, WEDNESDAY. At 7 am left and steamed up Haro Channel. As we rounded Turn point the weather changed as it always does; this point seems to be the turning p^{t} of wind and weather — dirty misty weather set in but no rain. We anchored at the S side P^{t} Roberts at 2 PM; *X* [cross bearing] of pt in one with Mt Constitution in 12 fms. Out paddlers and landed all the Lumber and gear for the Monument. Carpenters commenced to erect houses for Men & officers. Came off at 5.30 PM having visited the foundation for the Pedestal on the summit of the cliff 200 feet high. Thursday morning landed the carpenter to build houses, also 20 men under Lt Hankin and the Boats^{n} accompanied by Dr Campbell. At 10.30 Carpenters returned having got the houses nearly complete and left some people to finish them. Weighed and steamed for Nanaimo where we anchored at 3 PM. Evening set in rainy. Moored. Lying here the DM Hall and Retriever Merchant, ships loading with Coal for San Francisco. Friday, rainy weather. Easterly wind. Commenced coaling with our reduced crew.

NOVEMBER 16, SATURDAY. Coaling. Got 55 tons onb^{d} during the last 2 days.

NOVEMBER 17, SUNDAY. Cold weather, snow on Mt Benson.

NOVEMBER 18, MONDAY. Got no coals today. The place has not much improved as far as system goes. The[y] load lighters and let them sink for want of some one to look after them. Sent up the river for water, got 7 tons by night after the boat grounding and laying onshore several hours thro bad management.

Tuesday and Wednesday got a launch of coals about 50 tons. Tuesday rainy. Wednesday very fine. Attempted another load of water but got none & let the boat ground as before. Did not get her off till Thursday morning and then empty.

NANAIMO

NOVEMBER 21, THURSDAY. Weather set in rainy at 6 am. Baro up to 29.98. Unmoored and prepared to start. SE wind and rain continued but as glass remained nearly 30.00 I fancied it w^{d} clear, and weighed. On opening out the Strait it blew very hard and rained — so turned back &

anchored at 11 am. A schooner & barque which had sailed put back also. Blew and rained very hard and constantly all day, tho the glass remained steady.

NOVEMBER 22, FRIDAY. The SE gale continued strong throughout the day but the glass remained high, almost to 30.00. It rained the greater part of the day and poured in torrents all night. Fires banked since yesterday morning.

NOVEMBER 23, SATURDAY. Squalls and rain this morning, but the wind in the Strait having apparently dropped and Baro 29.90, I started for P^t Roberts to reprovision the party and see what they were about. It continued to rain all day and a fresh SSE wind was blowing. Shortly after 2 PM we anch^d on West side of P^t Roberts, Mt Constitution, on with *X*mc of spit or 9½ fms, ¾ mile from spit end. Found the Skow Schooner "Mary Anne" here having arrived this morning from N Westminster with the stones or 14 of them for the monument but, owing to the swell, had not been able to land them. Landed & saw the party — provis^d them for 10 days. Found the wind had been blowing here from South during the last 3 days, but not so strong as we had it at Nanaimo. From SE^d this place affords no shelter with the wind South^d of SE. Came off at 3.30, wind shifted to SSW, very dirty rainy evening, glass 29.80, but there seems no appearance of a gale and as it has been now blowing for 3 or 4 days, I should imagine we need not look for one.

<u>COAL MINES OF NANAIMO</u>. The following information has been gathered from Mr Nicol, Manager of the HB Coal mines. There are three seams now being worked. One at Newcastle I^d on "level free" — which is a very good kind of coal, but without sinking a shaft, there will not be more than 3000 tons available. The no 3 pit which is an inferior coal for steaming purposes, being very dirty with a large seam of fine clay running thro it — is being worked at 6 dollars a ton and there may be 40,000 tons to be got from the present shaft which is sunk 100 feet. This seam is 6 or 7 feet in thickness.

The 3^d seam is called Dunsmuirs — this is by far the best coal, very clean comparatively, and costs 7 dollars the ton. It is worked now level free and there is probably not more than 700 tons left on that level.

A new shaft is being sunk — over this field called the Douglas Vein. It will be completed about the end of January, and the yield calculated of it maybe 100,000 tons. This is problematical. This shaft is not intended as the main one, but to some present purposes — as the Co will not warrant the necessary outlay. The quantity of coal got out of the mines for the 12 months ending April, 61 has been 13,891 tons — and the 6 months ending October 61; 8,288 tons. The quantity shipped during the 12 months ending 30 April, 61 is 14,455 Tons, and the number of vessels taking it were 173. The greater number vessels of 25 to 70 tons — between Nanaimo & Victoria. During the last 6 months the demand from San Francisco has been great, owing to no shipments arriving from the Southern States.

The number of miners generally employed are 46, who produce upon an average from 2¼ to 2½ tons each daily. Their average earnings is about 12/6 [12 shillings, 6 pence] per diem. It should be added that the greater part of them are men who are habitual drunkards, and that they work as many hours of the day as suits them or as many days in the week. They are paid on the task work principle about 4/6 [4 shillings, 6 pence] a ton. The Companys Carpenters, Blacksmiths & labourers earn about 7/10 a day, but they are regularly employed and have lodgings, fuel & medical attendance gratis, this latter however is no criterion of the rate of artificers wages in the Colony. Moreover, I fancy they are not of the best description here.

There are two Clergymen, an Episcopal, a SPG[111] man and a Wesleyan. The latter has an Ebenezer.[112] There is no Episcopal church yet. Neither milk, butter or eggs can be bought at Nanaimo. Fresh meat can now be got — and bread. There is an Inn and a store besides the HB Co store, but few Indians permanently reside here.

[111] Society for the Propagation of the Gospel. The Anglican Reverend John Booth Good was assigned to Nanaimo between 1861 and 1867 (Fisher 1992: 137). During this time, he adopted the son of Skenahun, a Snuneymuxw head man. This son was baptized as Louis Augustus Good, and "was later to become a designated Chief of the band" (Littlefield 1995: 160).

[112] This is a term occasionally used by Methodists, Baptists, etc., as the name of a meeting house. Coincidentally, Ebenezer Robson was the name of a Methodist missionary who began work in Nanaimo in 1860 (Bowen 1987: 118).

§ 1861 P^{T} ROBERTS

NOVEMBER 24, SUNDAY. Last night there was a fresh breeze from SSE. Fell calm at 4 am, barometer down to 28.67, commenced to rise at 4 am, SW wind sprang up fresh. Weather cold.

*X*mc position at anchor with her stern swung in to a fresh SW wind was:

Rob Spit just outside Mt Constitution	S51E	6¾ fms at high water 5 at low f&c S with black [illegible] The anchor in 9½ fms About ¾ mile from Spit end
Flagstaff at northern end of flat	N64E	
Great Shingle ⚠	N9E	
The parallel	N11W	
Ships H^{d} SSW — dir^{n} 3° Westly		

NOVEMBER 25, MONDAY. A fresh wind from SW to SE blew last night. At 4 am it veered to NE, and was fine with very cold weather. The glass had been up to 30 inches, but fell more than 2 tenths. As the temp^{r} fell sent boats to tow the stone Skow on shore. Landed at 9. It was high water at 9.30, but the tide fell very little up to noon. We got the schooner cleared of her 14 stones in an hour and sent her back to Fraser River. She might have secured alongside the wharf for 3 hours easily. Got forenoon sights for time & sun latitude and true bearings am and pm to fix the line of the Monument. Wind blowing fresh from NW in the Strait and at 1 PM it reached us & set some ripple on the beach, but the bank protects the beach more from NW winds. With SW or west a surf gets up immediately. Got some Soundings at the Anchorage. Returned onb^{d} at 3 PM and hoisted up all boats; there is a Stream from which boats may water about ½ a mile northward of the parallel, but it is nearly 2 miles to pull to the safe anchorage for a ship.

NOVEMBER 26, TUESDAY. Weighed at 7 am from P^{t} Roberts for Nanaimo. Weather fine but cold. Wind SE a fresh breeze. Got some soundings outside Gabriola reef and carried a line along the coast to Entrance I^{d}. Anch^{d} at Nanaimo just after Noon.

The Barometer went down this evening rapidly and it rained very hard but no wind in the harb^{r}, tho we could see it was blowing outside from SE — and subsequently heard that at Victoria it was a heavy gale. The glass was down to 29.47, which is very low. It rose during night and in the morning the ground was covered with snow.

NOVEMBER 27, WEDNESDAY. Got water from the mill stream close to the old salt house by using 50 yards of canvas hose. If boats are sent to the mill they ground. Between this and Saturday we took in 36 tons of coal and filled with water. The weather very disagreeable — either snow or rain.

NOVEMBER 30, SATURDAY. Baro 29.80 and on the rise. Snow falling heavily. Weighed at 8 am and steered for P^{t} Roberts, in connexion with the boundary monument. Cleared with cold NE wind. Anch^{d} off the west side of the spit at 1.30 PM. Landed and laid off the east line for the Mason, Mr Thintson. The foundation had been laid — of granite masonry 10' 6" square. It lay in the true East and west direction and the parallel land off from the Boundary mark on the edge of the cliff. Ran thro the foundation at 4' 6" from south end instead of 5' 3" or ½ of the platform. Therefore altho it is hard square on the parallel the foundation is [illegible] 9 inches to the north of the line in parallel. In other words, it gives the US 9 inches too much territory if we build the monument in the centre of the platform which I thought it better to do for the sake of appearance.

Found our party had got the 14 stones on the cliff, but the schooner had not returned with the remainder. Returned onboard at 4 PM.

DECEMBER 1, SUNDAY. Snow and rain. Wind easterly. Got onboard 10 of our party from the shore, leaving 10 to go on with the work.

DECEMBER 3, TUESDAY. The barometer has been low, between 29.40 and 29.60 for the last 2 days and the weather has been bad with strong winds on Sunday night, a strong SE wind, and on Monday afternoon it blew so strong from SW that I was obliged to weigh and run into Semiahmoo bay where we anchored at 5 PM with the SW Bluff of the Bay in one with Mount Constitution bearing SSE — 2 miles from the Spit end — in 10 fathoms which is a very safe berth with plenty of room. It blew heavily in the night from west, but on Tuesday afternoon the glass rose and it moderated. Altho we have had constant bad weather, rain, snow and strong winds these last 3 weeks, it has never blown strong for many hours from one quarter and I suspect it does not in these inner waters. On Sunday the new moon 1 Decr we had high water at P^{t} Roberts — the superior tide at 8 am and the small tide about 2 PM.

DECEMBER 4, WEDNESDAY. Weighed at 7.30 am for P^{t} Roberts, thick fog till 10 when we anch^{d}. Landed and went on the cliff. Saw the Schooner on her way down with the Stones. Baro high 30.15 and signs of fine weather. Glass went down in 1^{t} watch to 29.99, fresh NE wind sprung up, increased at 4 am so much that I got steam. The schooner anchored astern of us during the night.

DECEMBER 5, THURSDAY. Glass falling & breeze from SE steadily increasing. Noon Baro 29.70; at 1 weighed, wind having increased rapidly. Much difficulty in rounding P^{t} Roberts. Blowing a very heavy gale. Got round the point by 4 PM & steered for Semiahmoo Bay. Blew away inner jib, and glad to get fore topsail off her. I have never seen it blow heavier in the Strait, with a greater sea. At 5 anch^{d} in Semiahmoo Bay, S^{o} Bluff being SSE in 11 fathoms. Wind 84 in Port and 40 on Strait. Bank^{d} fires. I fear it will go hard with the Schooner at P^{t} Roberts. At 8 PM Baro still going down 29.50, wind furious from SE. During the night it blew a whole gale from SE; anchorage well sheltered, rode easy. Baro went down at 2 am to 29.30, its lowest.

At 6am Friday 6th wind veered to SW, moderated; glass rising slowly. At 1 PM 29.54; wind moderated but w^{r} dirty looking. Weighed at 1 PM and steamed for P^{t} Roberts. Saw the Schooner safe at anchor but had dragged — as the weather looked very dirty and the Baro more inclined to fall than rise, I did not anchor on the western side of the spit, but in Boundary bay with Roberts Bluff and the spit end in one 1½ miles from the former in 12 fms. This with SW wind now blowing is as good anchorage as in Semiahmoo bay, and sheltered in a measure from west by the reefs off the Bluff. This gale has been a very heavy one and has lasted longer than usual — by 24 hours. Whenever a SE wind springs up with a high barometer — but a steadily falling one it lasts long & blows harder than when the glass falls rapidly — in like manner when the Baro rises slowly after it, the weather does not get fine immediately, tho there is more prospect of some continuance of better weather when it does come.

DECEMBER 7, SATURDAY. At 6.30 am weighed and steamed round the west side of P^{t} Roberts — anch^{d} there at 8 o'clock, sent boats & towed Schooner to the pier. Water very high, probably forced up by late SW gales. Got 18 stones out off vessel, 5 of them over 3 Tons weight — left 10 men under Lt Hankin to get them up the Cliff — which will be a heavy job — and at 12.30 weighed and started down the Strait for Esquimalt. We carried the Ebb till 2 PM — when about E. pt of Saturna got the flood. 6 days after new moon. The barometer does not rise above 29.65 but seems rather inclined to fall. Weather gloomy. The Strait full of trees & drift wood, washed off the high water line by the late gales.

1861 VICTORIA

Anchored at Esquimalt at 8 PM. The red light of Fisgard I^{d} did not shew well or as great a distance tonight as I had expected and a light on the shore (a fire) near Clover pt was at first mistaken for it. The safe way for vessels coming down Haro S^{t} is when Race Light is seen clear of Discovery I^{d} bearing SW or SW½S to steer for it until Fisgard I^{d} light is seen bright (when it will bear NW&W ⅓W); then steer for Fisgard taking care not to open out the red shade if intending to enter Esquimalt. If bound for Victoria and desiring to enter at night, which any steamers knowing the way can do — observe the same rule — steer for Race L^{t} until Fisgard I^{d} shows bright, & then steer for the latter until the light inside Victoria harb^{r} on Colville I^{d} bears north — & then steer for it which will take a vessel into the Entrance Midway between the points but as the entrance is narrow, the <u>north</u> bearing must be accurately got on and a look out kept as well. These precautions render it necessary for a vessel to make a somewhat circuitous course by which a mile or two will be lost — but they are the only safe ones to avoid Trial I^{d} and the points east of Victoria, including Brotchy ledge. A vessel must be careful coming down Haro S^{t} not to steer for Race L^{t} as soon as it is seen or it may take her onto Discovery I^{d} — when the L^{t} bears SW^{d} S, it is in a line with the E end of Discovery I^{d} so that it must be brought to bear SW½S at least, or better SW before steered for.

DECEMBER 8, SUNDAY. It rained the whole day. Monday was little better. Tuesday morning weighed and steamed into Victoria harb^{r} just before HW. Neap tides carried no less than 17 feet in the Channel, but the tide was within a foot of its maximum at Springs and we were in the centre of the narrow channel. A spar or tide pole is set of[f] at McCauly pt — and a reference marked 18 feet noted on it, signifying that there is 18 feet on the bar when the water is up to it. It is not so, however, but 2 feet less. I believe Mr Pemberton put this mark up — wrongly.

It rained unceasingly until Saturday night and is raining now. Little wind for the week but SEly. A freight ship "Prince of the Seas" anch[d] in Royal Roads on Wednesday, and has been detained there ever since. Such is one of the drawbacks of Victoria harb[r] and cases of the same kind are constantly occurring.

> DECEMBER 10, TUESDAY. Surveying Officers employed on board in the Chart Room preparing the sheets for Tracing and transmission to England; the work this year has been plotted as follows: the whole of Barclay Sound & the Alberni Canal — on 3 in to the Nautic Mile — taking 5 or 6 Sheets of Double Elephant to show the whole — from them it has been reduced into an inch — Clayoquat Sound on 3 in — and plans of Harbours, anchorages, etc. into 4 and 6 inch to the Nautic Mile (Gowlland 1861–1864).

DECEMBER 26, THURSDAY. For the last 10 days the weather has been very fine, cold and frosty. Therm[o] standing below 30° at night. The glass at the commencement of this weather stood at 30.60, it has gradually gone down, but still remains cold and fine. Victoria harb[r] partially frozen over. The Prince of the Seas and Pruth, English ships entered the harb[r].

Had 2 meetings of the harb[r] commiss[n]. Light house stores arrived by the Pruth, or rather Prince of the Seas. Diving apparatus for Squadron by Pruth. Old M[r] Work, Factor of HB Co, died on 22[d]. Treasurer of the Colony, Cap Gordon,[113] arrested for defalcation of public money on 24th.

---

[113] G. Tomline Gordon, colonial treasurer.

▲ Left to Right: H.M.S. *Termagant*, H.M.S. *Alert* and H.M.S. *Plumper*, 1860.
(ROYAL B.C. MUSEUM & ARCHIVES: A-00239)

▲ *Hecate* on the Rocks, c. 1862. (NATIONAL MARITIME MUSEUM, GREENWICH: PX-9919)

▲ H.M.S. *Plumper* in Port Harvey, Johnstone Strait, c. 1860. Engraving of drawing by Edward Parker Bedwell printed in the *Illustrated London News*, March 1, 1862. (ORMSBY, 1958)

▲ H.M.S. *Hecate*, Esquimalt, c. 1862, watercolour by Edward Parker Bedwell. (ROYAL B.C. MUSEUM & ARCHIVES: PDP05357)

▲ Esquimalt, c. 1860, watercolour by Richard Frederick Britten, midshipman on H.M.S. *Topaze*. (ROYAL B.C. MUSEUM & ARCHIVES: PDP05446)

▲ A Street in Victoria. (ROYAL B.C. MUSEUM & ARCHIVES: PDP01892)

▲ Victoria, c. 1860, watercolour by R.F. Britten.
(ROYAL B.C. MUSEUM & ARCHIVES: PDP05437)

▲ Race Rock Lighthouse, Vancouver Island, c. 1860, watercolour by R.F. Britten.
(ROYAL B.C. MUSEUM & ARCHIVES: PDP05437)

1862
Survey Season

JANUARY 9. From the 15 of Dec[r] till the 6 of January — there has been a succession of fine, cold, clear weather with northerly & NE winds. The lowest the thermometer has stood at, being 18°. The Fraser river has been frozen over below N Westminster — and vessels have been considerably injured by the field of ice grinding into their sides and tearing their copper off; for several days port vessels have not been able to enter the river. On the 6th Jany Snow fell — and has continued up to this time today. It lies a foot on the ground and numerous sleighs have been driving about the Town & Country.

Parsons came down on the 6th. D[r] Lyall arrived from [illegible] & Colville on 29 Dec[r] by "Brother Johnathan" & was invalided — here.— returns to England. General steamer the Sierra Nevada arrived from San Francisco on the 3[d] & sailed [a]gain Same day. The thermo at N Westminster has been down during the late cold weather to 9°.

> JANUARY 13. Great excitement is prevailing in the town at present, caused by the discovery of large frauds having taken place in the Treasury with the Public Money — to the amount of 1200£ nearly. The Treasurer Wm George Tolmie Gordon, ex Captn in the Light Dragoon Guards; and his clerk; were arrested and imprisoned on suspicion of the thefts — it would doubtless never have transpired had not His Excellency, suddenly, without giving any warning to the public officers of the Government caused the accounts to be audited and before the Treasurer had time to refund the money fraudulently taken, his deficiencies were discovered. The clerk after been kept in prison for one Month obtained his discharge on the Grand Jury finding no bill against him. Mr Gordon was tried for embezzlement, but through the stupidity or ignorance of the Attorney General, he was acquitted — the Counsel for the Prisoner Mr King having discovered a flaw in the indictment and the Jury returned a Verdict of "Not Guilty" alltho their inclinations and sympathy were against the prisoner, he was again arrested on a new charge but subsequently made his escape from jail and gained the American shore before he could be caught. This is the 2nd Gov^{t} official who has betrayed his trust and robbed the Colony during the last 6 months (Gowlland 1861–1864).

JANUARY 21. Snow has covered the ground since the 6th and the temperature has been very low — down to 9°+ at night on 2 or 3 nights. On the 11th the Brother Johnathon steamer arrived from San Francisco and left same day. Dr Lyall went in her. She brings a rumour of the death of Prince Albert by telegram across the Continent.

JANUARY 22. Cortes steamer arrived from San Francisco — brings news of Mason & Slidell having been given up by the U.S. Govt. and an apology made to ours.[114] Also that Brother Johnathon, which left here on 11th, was still frozen in 60 miles up Columbia. Cortes left again on 23^{d}.

Communication still cut off with Fraser River by ice. A small st^{r} the Emily Harris, left for Burrard Inlet on the 20th in order to communicate and get provisions across the trail to N Westminster. On the 21 Jany there was a partial thaw — but on the following day snow again fell, and likewise the temperature. On the 25th the harb^{r} was frozen over so that with difficulty we could get a boat on shore. There has also been a difficulty about getting fresh water for some time past. Indeed, so severe a winter has not been remembered here.

We have been employed getting the deck & boat repaired and refitted, ready for service on the West Coast, and painting and putting to right the Gigs, ready for work, but this severe weather has been a great drawback. The Forward gunboat was hauled up on a slip here on the 12th to give her a new keel & copper.

FEBRUARY 7. Since the 30 Jany there has been a rise in the Temperature tho it has been below the freezing p^{t} generally at night. Snow has not disappeared from the ground since 6th Jany. The Cattle are suffering very much, and considerable quantities come over from Oregon by steamers, tho I don't think any considerable increase over former imports. The Princess Royal HB Co ship arrived outside on 3^{d} of Feb and came in on the morning of the 4th, having laid aground in the entrance of the harb^{r} all night. The only communication with Fraser river up to this time is by Burrard Inlet by a small steamer. Cap Darrah of Boundary Commiss^{n} arrived on 3^{d} from Semiahmoo and left again on 6th in Grappler with provisions for his people, and to put the pillars up on P^{t} Roberts isthmus. On the morning of the 8th Feb "Charly", an Indian, was hanged for the murder of one of the Topaze men. He observed just before he went off that he should die happy if "Tom", who took a more active part than himself, were also hanged.[115]

[114] In November, 1861, James Mason and John Slidell, representatives of the southern Confederacy, were en route to Britain and France to request formal recognition as an independent nation when their ship, the *Trent*, a British Mail Packet, was seized by Union forces and Mason and Slidell were arrested. The British government called the Trent Affair "an affront to the national honor" and demanded the release of the prisoners and an apology. Mason and Slidell were freed on December 26, 1861, but no apology was made (Goodwin 2006: 396–401).

The whole of the month of February the weather continued severe. Temperature below freezing point every night, and rising 6° above in day time. Snow fell in quantities — but not much strong wind, it being principally from the north[d], and light. Towards the end of the month there were some fresh SWly winds.

On 22[d] Feby we launched the Shark, having been waiting nearly 3 weeks for a tide high enough. The northerly winds had kept the water very low.

MARCH 1862. The severe weather continues at the end of the 1[t] week in this month. On the 3[d] a thin coating of ice formed in the Shallow portions of the harbour.

Employed preparing for a Start. Boats and Surveying gear undergoing repair and refit.

MARCH 8. The therm[o] down to 24+.

MARCH 10. The HB Co ship "Princess Royal" sailed. She grounded going out but got off in ½ an hour. The Shark was dispatched today on surveying duty — and to meet us at Nanaimo.

1862 VICTORIA–NANAIMO

MARCH 15, SATURDAY. Unmoored this morning. Weather very dirty looking, but strong SW wind set in at 10. At 11 am the water having risen sufficiently high, we weighed and steamed out of harb[r]. Met a barque under sail in the narrowest part of the Channel and by very close shaving kept clear of her on one side and the ground on the other. Entered Haro S[t]. Strong SW wind and Squalls of rain. Made Sail. Entered Plumper Sound at 2.30 PM and anchored at 3 in Browning harb[r], which is a very snug good anchorage. Blowing very hard outside. Walked across to Bedwell Harb[r], only separated from this one by a narrow portage of a few yards. Came off at 4.30 PM.

---

[115] Thomas Holmes, a stoker on the *Topaze*, was murdered in Victoria on December 8, 1861. Holmes' body was found at 10 a.m. on December 9. Two Cowichan men, Tom and Charley (a.k.a. Klorek) were arrested by noon and at 5 p.m. the same day, the inquest began. Evidence at the inquest showed that Holmes went to a "dance" at the big house and began yelling. He was later found dead. Both men who were arrested claimed that the other was lying. Both were found guilty (*British Colonist* 1859–1862: December 10, 1861, December 12, 1861, and February 8, 1862).

MARCH 16, SUNDAY. Day after full moon. After Church weighed, steamed thro Navy Channel and into Plumper Pass. It was just past low water and we carried 2 or 3 knts flood thro the pass at 20 minutes after Noon. It appears slack low water in the pass on F&C days at 11.30 am — which is the best time for a ship to pass thro.

Shaped a course to pass outside Gabriola Reefs — at 3.50 passed Entrance I^{d} and anchored in Nanaimo harbour at 4.50 PM. Snow covering the ground and everything looking wintry. Remained at Nanaimo until the 21^{t} Friday, and during our stay got observations for time, T Bearing and Magnetic, dip etc. Were unable to swing the ship on account of bad weather the day before we left, so postponed it till our return, as also coaling, as it was my intention to go up the Fraser River and it is better to go light.

1862 NANAIMO–FRASER RIVER

MARCH 21, FRIDAY. Weighed at 10 am. A thick fog came on just as we left the harb^{r} and I turned to drop an anchor on the edge of the South bank outside Gallows point — dropped the anchor in 8 fms or rather intended to, but it hung & she was in 4 immediately, her forefront taking the ground with 5 fms aloft. Tide falling, ran guns & shot aft. Turned astern. All hands aft & in 10 minutes she came off. Steamed out — fog clearing a little — very dirty day. At 2.30 saw S^{o} Sand hd Buoy of Fraser River — ran in NNE between it & N^{o} Sand buoy, keeping a cable off the latter and anch^{d} in 20 feet just within; then sent boats to ascertain what buoys remained. Found 3 inside the N^{o} Sand H^{d}, but only 12 feet on the shoal flat part of the bar. So remained at anchor. At 5.30 "Otter" passed in; boarded her & got newspapers.

MARCH 22, SATURDAY. Very foggy, but water appeared high. Weighed and steamed in on a NNE course. Got a glimpse of 1^{t} buoy, went on and made 2^{d}. Cleared a little. 3^{d} buoy had got out of the channel to the South. Passed 1½ cables south of it. Entered River at 8, having had 18 feet least water. Anchored off NW camp in thick fog at 10.15 am; found ourselves between Sand Bank and Shore. Shifted further down, anch^{d} in 9 fms. This day turned out hot and fine. Got PM Sights and arranged for the lithographing of Barclay Sound, my principal object in coming here.

Col Moody absent. Remained here until Friday 28 March. Constant rain since the preceding Sunday, and no ob^{r} beyond those I got the first day, except Magnetic. The Ebb Stream has not run between the 17th & 28th more than 3 kts. Seldom so much. Ship Swung to flood for a few hours (5) during night.

MARCH 28, FRIDAY. One day before new moon. High w at about 6 am at Camp, and about 4 PM. Started at 5 am, swung to flood, at Garry L^{t} before 8, and Steamed thro Shoals by 8.40. Least water 18 feet. Strong breeze up the Gulf; made Sails and anchored at Nanaimo at 11.30 am. Raining and dirty w^{r} latter part of day.

Friday and Saturday — Weather bad. Commenced coaling as soon as we anch^{d}, and had 61 tons in by Saturday Eve. Found our decked boat the Shark here on our arrival.

The Episcopal Church[116] is fast progressing towards completion and is a very good specimen of Church architecture — a great contrast to the hideous Ebenezer.

Remained here until Monday the 7th April, coaling and getting observations Magnetic and for rates etc. Swung the ship for local deviation, took in nearly 200 tons of coal — a mixture of Newcastle, Dunsmuir, and No 3 pit. The 1^{t} and last at 6 dollars a ton, the Dunsmuir at 7$. The Newcastle turned out rubbish.

1862 NANAIMO–SEYMOUR NARROWS APRIL

On Saturday the 5th April we rec^{d} our mail. On Sunday secured equal altitudes and on Monday at 4.30 am started for the North. The day was fine and something like spring. Calm weather. Passed Ballinacs (inside) at 7.30 am. These I^{ds} about 200 feet high, NE one bare and peaked; one or two trees on summit; SW one wooded. They appear as one I^{d}. Passed East^{d} of Hornby I^{d} and 4 miles from Cape Lazo. Abreast latter at noon. We carried a flood or northerly Stream with us until between Lazo and Mudge, where the 2 tides meet and found the Ebb helping us north^{d} at

[116] The first St. Paul's Anglican Church opened in June 1862. It was the first of three built on the same site (Sale 1986: 7).

this spot. It is 7 days after new moon. Consequently neaps. No tide rips abreast of Cape Mudge at 3 PM, the ebb about 2 kts with us. Saw no Indians at the old Villages. At 4 PM we were at Seymour Narrows. It was just slack water, or a little ebb running but appeared to have been low water by the shore perhaps more than an hour. What favours my belief that the Ebb stream of the Narrows runs after low water — we anchored in Plumper Bay 1½ miles north of the Narrows at 4.30 PM. The flood or southerly stream seemed just to be making. Anchored in 9 fms in the Centre of the bay about 2 cables from the shore. A very comfortable stopping place out of the tide. It appears to be low water here about 4 PM at Neaps — and certainly 4 PM and 10 am is the best time to pass thro at this state of the Moon.

From former observations — it is high water (springs) about 5 o'clock. Hauled the seine with no success. Sent to examine the beach line of the bay and found as I suspected that there had been an error in plotting its head. Our survey makes the bay too deep and shews 9 fathoms — 3 cables lengths from the shore, whereas that depth is obtained at a little over one cable only. This detracts from the value of the place for sailing vessels, tho not for a steamer. It is perfectly safe, out of wind or tide, and one of the best stopping places in the Straits — far better than Knox Bay; but it ought not to have been so much in error in the chart; the result of hurry and inexperienced hands, but probably most of all my fault. Corrected it.

APRIL 1862 JOHNSTONE STRAIT–FORT RUPERT

APRIL 8, TUESDAY. The morning very dark, raining and dirty. Baro fallen nearly an inch (29.25), but as our fine and foul weather seems to occur on alternate days, and as I wanted a fine one at Rupert and am averse generally to delay, we weighed before 5 am. Found at that hour some 2 or 3 knots of flood or southerly tide against us, but kept the eastern shore on W and when abreast deep water Bay 2½ miles above the Narrows, were out of its influence — and were not at all inconvenienced the remainder of the day. Indeed, except at the Seymour Narrows — or in the passages between Helmcken I^{d} and the two shores of I^{d} & Main, there is little tide to inconvenience a ship during neaps. At 9 am we passed Helmcken I^{d} — little tide. It seemed about high water (neaps), which proves that the tide is earlier to the West^{d} as we found on former occasions. Thus HW at Helmcken I^{d} 28 miles west of Narrows an hour or an hour and a half earlier (the flood being from the NW). It rained or snowed and was very thick all day, with light SEly wind. Land obscured, but navigation very easy. Passed Port Harvey at 11(about) am — there is good anchorage.

Forward Bight a mile or 2 west of Harvey, well shelt^{d} from NW winds and a good cove for Schooners with 2 fms of water. At 3 PM passed Malcolm I^{d}. Saw the Shark at anchor on the main shore of V.I. 9 miles from Rupert. Signalled her to join us when weather permitted. A strong NW wind had just set in — which would have prevented her doing so then. Barom^{r} began to rise as soon as the wind sprung up. At 4.50 Anch^{d} at Beaver harb^{r}, Fort Rupert. Have not seen a Native or a Canoe since leaving Nanaimo — are apparently shut up in winter quarters and everything looking wintry enough certainly. Anchorage Cormorant Rk between Peel pt — 2 cables dist from Shell I^{d} in 10 fms.

APRIL 8, TUESDAY. Since our last Visit to this place in 1860 on H.M.S. "Plumper" great changes have taken place amongst the Indians; the population has decreased to about 300 men & women; numbers of them have left for Victoria, and all the young girls remained there living with miners; numbers of the men have hired themselves out as labourers and house servants to settlers, and given up their savage ways and Mode of living and adopted the modes of Civilation. Numbers of the tribe of both sexes old & young; have after a lengthened visit to Victoria just managed to crawl back to die; from the affects of Disease incurred by their connection with white men; so that the place has degenerated in more ways than one. Old Whale the 2nd Chief died some 3 months ago of some internal complaint; Jim our old friend of '60 paid us a visit and expressed a great desire to accompany the Ship on the West Coast till at last the Captain assented to his wish. Another great evil that has visited these poor wretches lately is Whiskey; some Americans fitted out a little Schooner, and loaded her with Alcohol and Camphene;

> and started for a trading cruize to the North; touching at all the Indian Establishments on the road; Fort Rupert amongst others; when they bought up all the Skins and blankets for this abominable stuff, causing quarrelling, Murders and some most frightful scenes in the Camp, after they had imbibed a sufficient quantity — and here these rascals take up their residence; keeping the Indians constantly supplied; and consequently always in a state of maddening drunkenness; cheating the honest fur traders out of their lawful traffic — and ruining the poor indian body and soul by their illicit trafic — a great shame the Government does not put a stop to these rascally proceedings (Gowlland 1861–1864).

APRIL 9, WEDNESDAY. Strong NW wind all day and cold. Got equal alts for rates. Shark anch^{d} in the eve. Leaking in upper works. Very few natives here, are absent on a feast or drinking bout at Malillacolla.[117] Some fishing for Hulicans at Knights Inlet. 2 white men here selling spirits. Mr Moffat HB Co officer in charge of Fort tells me Old Whale (Chief), died 6 months since. Cut 2 logs of yellow cypress.

APRIL 10, THURSDAY. SE and East wind freshened gradually all day, ending in an uncomfortable swell. At night blew hard. Let go 2^{d} anchor. There is shelter with this wind under Shell and Cattle I^{d}. Surf on beach renders it difficult to land with this wind. A very bad rocky landing at all times. Very bad for boats bottoms. Provisioned and patched up Shark. No boats down.

APRIL 11, FRIDAY. Heavy easterly wind and constant rain all day. No boats down. One or two small canoes off with women selling a few clams. They have umbrellas. Found the tides here agree with our last observations. HW on F&C about noon. All thro Discovery Passage & Johnstone Strait the Ebb or northerly stream commences to run about 2 hours before HW and runs for 2 hours after low by the Shore. At Seymour Narrows it is hw on F&C at 4 PM.

---

[117] Mamalilikulla. The main village of the Mamalilikulla is on Village Island, south of Fort Rupert.

Flood makes about 11.30 am and runs until about 5.30, or 1½ to 2 hours after hw.

⸹ FORT RUPERT

Fort Rupert was built in 1849 by Cap Mc Neal [McNeill], one of the HB Co Factors now stationed at Simpson. It was for years one of the chief trading posts on the Coast for Furs, dogfish oil, Salt Salmon. Since the trade in furs etc. has been thrown open and Victoria has sprung up, the Indians find a better market for their skins there and, as there are also other sources of profit less creditable, they resort there in great numbers with their young women — and little trade is now carried on here. The fort tho in outward appearance sound enough, is much decayed and will probably soon be abandoned. The Indians in appearance are superior to the Southern tribes which have come since in contact with civilizing influences, but the introduction of spirits into this part of the I^d by adventurers who trade in that article for skins is working its ruinous effects rapidly on the people. There are 4 tribes at Rupert. They used formerly to live at separate villages, but have been attracted here by the nearness of the fort, and the Village is the most extensive one I have seen. The names of the Tribes are: Quee-peer; Quah-quolth; Kom-kutus or "Wah-lish Quah-quolth"; and the Loch-quah-lillas.[118] The Nimpkish tribe number 100. They live at the river of that name about 18 miles from here on the East Coast of the I^d. At Cape Mudge at the entrance of Discovery Channel there are also 4 Tribes, amounting to about 500.[119] They are migratory, never remaining long in one place but during the Hulican and Salmon seasons removing to the scattered villages & stations in the deep inlets of B.C. There is a tribe of 100 called "Mort-teelth-pat" in Call Creek. In Knights Inlet 2 tribes,[120] 50 in each. Two in Broughton Archipelago, the Kloitzers and the Mah-mah-lillicullas, the

---

118 Kweeha or Kwiakah, Kwakiutl, Komkiutis or Walas Kwakiutl, and Lackwilalas.

119 Laich-Kwil-Tach included the Wewaikai, Weiwaikum, Kwiakah at this time. The fourth group may be the Tlaaluis who, at about this time, joined the Kwiakah, the Awahoo who amalgamated with the Weiwaikum, or the Walitsama who were living at Salmon River.

120 Likely Da'naxda'xw and Awaetlala.

former 120, latter 250. In Fife Sound 2 tribes,[121] counting 160 between them. A very small tribe of 10 in Wells Passage.[122] On Galiano I^{d}, Goletas Channel, a tribe of 180 called Klattle-se-Koolahs. At a village called Na-whitti, also a tribe of 150, the "Nah-kok-toks" who come over from the main during the Hallibut season and put up at Nawhitti. A small tribe, almost extinct not more than 6, reside at Sea Otter Bay, "Ne-Kim-Kle-sellers." After these come the Quatsino Sound people. The Rupert language is very extensively spoken, entirely so between Cape Mudge at the South to Cape Caution on the main, both on the Island and Continental shores and as far round as Port Brooks on the west side of the I^{d}. There appear to be 4 principal dialects, or rather distinct languages — the Songhies between Sooke and Cowitchin; then the Nanaimo to C Mudge including Fraser River; the Rupert; and the Barclay Sound which latter is spoken from C Beale to Nootka with slight variations.[123] The Hulican season is early and nearly all the Natives of this place are absent at Knights Inlet taking them. They are found in peak abundance at the heads of these great arms, are very delicate like the Smelt, and the Natives dry them and make them their principal article of food till the Salmon come in, in August. The Hallibut season is later — in September and October — at least the fish are better then, tho they are caught all the year round.

SHUSHARTIE–CAPE SCOTT

APRIL 12, SATURDAY. The SE wind died away last night, but baro fell this morning and it came on as fast as ever with thick dirty weather — all yesterday it rained incessantly, with cold miserable weather. Weighed the 2^{d} anchor this morning and sent Shark to Sushartie to wait fine w^{r} for the outside coast. I am waiting here for observations for rates. Got 2 logs of yellow cypress off during the lull this morning. It blew fresher towards evening and the 2^{d} Anchor was let go again.

---

[121] Tsawataineuk and Qwe'Qwa'Sot'Em.

[122] Likely Gwawaenuk.

[123] The language groups today are known as Straits Salish, Halkomelem, Comox, Kwakiutl (Kwak'wala) and Nootka (Nuu-chah-nulth) (Thompson and Kinkade 1990: 32).

APRIL 13, SUNDAY. Weather not improved.

APRIL 14, MONDAY. I succeeded in getting observations. HW at noon. Full moon at 2 hours today. At 2.30 PM left for Suchartie, steamed thro Goletas Channel and anch^{d} at 6.15 PM. Peaked rocks between Mt Lemon a cable from the eastern shore in 8 fms very low water. The Shark came in at 8 PM and Mr Blunden reported her very leaky. I have decided not to risk her on the West Coast in her present state — indeed under any circumstances she would have been cause of considerable anxiety to me, so at 5 am I sent Mr Bedwell in her to Esquimalt at 10 minutes warning, with a request to the Land Officer to ascertain the leak and repair her. M^{r} B to work in the Gig left behind at Victoria about the East Point of the I^{d} and Discovery Channel.

APRIL 1862 CAPE SCOTT TO QUATSINO SOUND

APRIL 15, TUESDAY. Left Suchartie at 5.30 — nearly low water. Steamed over Nahwitti bar. Swell heavy but no wind, and a heavy surf all along the Coast. Dropped Mr Browning in 3^{d} Whaler to examine the shore to Cape Scott — and stood northward to Sound. Found a bank of Gravel extending for 2 or 3 miles west of Hope I^{d} with as little as 7 fathoms on it. Carried a line of Soundings westerly towards Cape Scott — which we passed at noon. NW wind sprung up and weather appeared unsettled, tho Barometer above 30 inches. Cape Scott is a promontory, rocky, covered with trees 150 feet high connected to the I^{d} by a sandy neck apparently narrow. 2 or 3 rocks above water extend about 2 cables west^{d} of the pitch of the Cape, the Channel between it and Scott I^{d} is very good. Steered along shore to the South^{d} about 2 miles off. The first prominent point south^{d} of the Cape and tangent from it is the Northern pt of Sanjose Bay [San Josef Bay], a rocky I^{d} 40 or 50 feet high lies off it. Rounding this islet about a mile and standing into San Jose Bay, the entrance to what is shown as Sea Otter Harb^{r} on the chart is seen. It appeared very narrow — and a reef extending ½ way across it at h water — sent boat in to examine the entrance, which returned reporting no water for the ship. Kelp right across the entrance which is not 50 yards wide, and a rock with 13 feet at h water in the Center. There was a heavy surf all along the Coast and a breaker extended for nearly a mile off the Rocky I^{d}. San Jose Bay did not appear to offer any protection, so

I steered out for Quatsino, dist 17 miles. Off the south pt of San Jose breakers extended for ½ a mile or more. Sounded at the distance of 1 or 1½ miles — in 30 to 20 fms gravel. Entered Quatsino at 6 PM. Both Danger Rocks, were breaking — it being near low water and much swell. Entering Quatsino from the North, it is better to pass between the Entrance I^{d} and the N^{o} Danger rock, keeping 2½ cables off the former. The hills over the NW entrance pt are rather remarkable, and Entrance I^{d} small, wooded and 80 feet high stands a cable off the pt. A reef extends 150 y^{ds} off the I^{d} pt. Steer in NE, keeping 2 to 3 cables off the Id. As soon as Pinnacle I^{d} (a split I^{d}) is in one with the point west of it, or with the pt of Low I^{ds}, steer up to pass Pinnacle 2 cables. North Danger will be ahead the same distance on the starb^{d} bow. Pinnacle I^{d} is rocky, 30 feet high, with a few trees on its Summit which from the south^{d} make as a top knot or crest. Low I^{ds} appear as one — are wooded, the tops of the trees about 70 feet high. The hill over the eastern side of the outer harb^{r} is remarkable, 1000 feet high about. The northern half bare of pine trees from fire, the bushy alder has taken their place & forms a remarkable contrast. After passing Low I^{ds}, round Robson I^{d} about same distance and take up a berth between it and north shore in 6 fms. Although an excellent harbour, the north harb^{r}[124] of Quatsino and very convenient, being so near the entrance, the Danger rocks make the Entrance awkward for a stranger. In a sailing vessel particularly, if running in with a strong wind which makes much sea, but at the same time causes the rocks to show. Entering from the south^{d} a ship should pass south^{d} of both rocks.

APRIL 1862 QUATSINO SOUND

I remained on the 16th at Quatsino & got observations, and on the morning of 17th April at 5 o'clock, the Baro being 30.50, left again with the intention of taking 2 boats to Sea Otter Creek and leaving them there also to get observations for time and Lat etc.

We had no sooner got round Entrance I^{d} than the wind sprung up from SE, which I congratulated myself on, as the Barometer was very high

[124] North Harbour, located to the northwest of Robson Island (now Matthews Island), is a bight on the west side of Forward Inlet (*BC Pilot*: 340).

30.50. I considered fine weather certain, and was rather pleased at the absence of our Old Enemy and surf maker NW— but it soon freshened so much bringing with it a heavy swell and dirty weather — dispelling all hopes of getting observations or being even able to land my parties at Sea Otter Cove that I began to feel anxious. At 7.30 it was a gale — the Barom^r steady at 30.50, even rising. There were only 2 courses open — stand off under sail, or endeavour to fetch our old anchorage. The latter was far preferable — all our canvas was rotten and smoke dried, and burning coals was objectionable. We therefore put her head to the SE^d and she steamed very fairly against the Strong wind 3½ kts on 1^t grade. The sea heavy and knocking about considerably. We anchored in North harb^r about 10.30 am and here there was no indication of a gale outside save the slight break on the beach on the opposite or eastern shore of the harbour. The glass kept high all day, but the weather was cloudy with rain. I have come to the conclusion that no settled weather can be looked for on this coast till the middle of May. It was so last year. I came out early this season hoping to be able to secure some of the Coast by the boats before the regular NW winds had set in, but I believe I have been mistaken. There seems a heavier swell on the shore than there is during summer, with the NW day winds and the land winds of night — which more or less prevail, then the fact is that during winter and till May, Gales are the rule — and strong winds from any quarter send a surf on the coast and make the shore treacherous for boats. It is under any circumstances and at all times an uncertain coast, and calculated to turn ones hair grey or white — and to inculcate patience.

This evening our old Indian, Rupert Jim, informed me that the girl with the Long head had come down to see me. I have mentioned this girl on our last visit as having a most remarkable conical head. I asked for her on our arrival, and it appears that they sent for her 17 miles off at a Village which they frequent at this time.[125] She came down with her father and mother, and 2 or 3 old Indian friends of ours. She had grown a good deal and was far less timid — came on board immediately, instead

[125] Probably Quattishe or Hwates, the winter village of the Koskimo, located at Hecate Cove in Quatsino Narrows. In the 1860s there were fifteen houses at Quattishe, now Quattishe IR#1 (Galois 1994: 370).

of making objections as she did in the Plumper. I gave her some rigging and ornaments and professed to take a great interest in her, which I did and shewed her the original drawing of her head which she rather repudiated. She evidently saw she was a favourite. Her parents appear to look upon her as a prize, since the attention that has been paid her by Plumper & Hecate. She is now between 11 and 12 years old — her name Yak-y-koss.

Her father and mother appear intelligent people — none of them expressed the least surprise at our having changed from the Plumper to the Hecate, although the ships are so dissimilar. I imagine they look upon it just like changing canoes with themselves — but to shew that they are not wanting in observation, they remarked immediately — who were the old officers of Plumper and picked out Mr Hand the 1 Lt who was not here in her as a stranger before they had been onbd 2 minutes. This is singular — seeing that we were here more than 18 months ago.

QUATSINO SOUND–SAN JOSEF BAY

APRIL 18, FRIDAY. Blowing heavily from SE with gloomy weather. Pulled to the head of the western creek[126] and walked across to the outside coast — which is not more than ¾ of a mile with an Indian trail. Came out on a Sandy beach about 4 miles NWd of the entrance of this sound. Found that by walking to a point about a mile to the NW I should be able to get a station for cutting off the points to the westd and perhaps the Islands. Returned onboard at 1 PM — a boat employed marking the harbr for a larger plan I intend to make.

APRIL 1862 QUATSINO SOUND

APRIL 19, SATURDAY. A thick fog this morning which prevented my leaving at 5 o'clock as I had intended. At 7 we went out and steamed to the NW sounding and fixing points and hills — a very heavy swell but fine and little wind. At 11 we were abreast San Josef Bay and I anchored about ½ a mile off the entrance of Sea Otter Cove in 14 fathoms sand while I got a Latitude and afternoon observation for time and true bearing. The anchorage not comfortable, but answered our purpose for a

[126] Probably Browning Creek.

few hours as a stopping place. At noon the White Whaler, Mr Browning, rejoined from Cape Scott, having had a rough cruise — and found no kind of shelter for a vessel between Suchartie and the Cape — with difficulty and risk could beach his boat.

Here, at Sea Otter Cove, I left 2 whale boats with Mr Pender the Master and Mr Gowlland, 2^d Master, the former to examine the Scott Islands, and the coast between Cape Scott and this place; the latter to survey San Josef bay and Sea Otter Cove, and to examine the coast down as far as Quatsino. The Entrance into Sea Otter Cove is very narrow — only sufficient to enable a Gun boat to enter when within, a very small place but smooth water and perfect shelter. At 3 PM, having obtained my observations, I returned to Quatsino, carrying a line of Soundings down, and anchoring in our old berth at 7 PM.

APRIL 20. SUNDAY. Easter day. The barometer high 30.50, and sun out but cold and windy. Got observations for time and latitude.

My long headed little Indian girl and her friends left us today for the upper part of the Sound. I don't think they realized as much as they expected, but the fact is we are very ill provided with presents for natives and after all, the clothing of the child in yellow flannel and blue serge, with a Crinoline manufactured by the Cooper, was no great remuneration for bringing her 20 miles.

Blankets are the only things of real use to the Indians, and they can scarcely comprehend our coming here without trading. Another long head was brought off in the afternoon, whose friends evidently expected was to be clothed also — but she did not make the same impression as our young friend, being anything but pretty, and passing old likewise. I'm afraid our admiration of the cones may induce a continuation of the practice of distorting the children — it does not seem to be very general and not nearly so prominent in the grown people as the children. I imagine they grow out of it a good deal. We have not got much fish here yet, no Cod or Salmon. Hallibut and a kind of Ling or Hake are abundant enough. No Deer. The Geese have gone north. There are a few here, but very wary. Ducks and the American Robin pretty numerous, but the Stock or red legged Duck is slightly fishy even on this coast. The Red fish is also rather abundant. It is considered very good but I think rather coarse. The men consume all they can get; indeed, their great object seems to be to save their provisions and get paid for them — Sailors even are getting sordid in this age. The Natives have pretty well left as I suppose they find we are not very remunerative. A few of them bring fish. The greater part are at the Upper Villages, and most of the men they say are at the Scott I^{ds} fishing for Hallibut.

APRIL 21, MONDAY. I walked with my boats crew and Hankin across to the west coast — with the view of getting on a point on the outside which would see something of the Coast and I^{ds}. The trail to the sea was passable, but the water being rather high, we could not get along the shore to the NW, but had to strike inland where I have rarely found such walking in my life: on the tops of banks or along fallen logs, and the bush so thick as to entirely puzzle one. After an hour or twos' scrambling, we gained a few hundred yards and again came out on the coast scratched and torn and done up — and did not get near the point. Fortunately for our reputation, it came on to rain & blow from SE so that it would have been useless proceeding. We returned by the Rocks and I determined the next time I tried I would study the tide if possible. But this is a difficult country, one moderately fine day in a week at the utmost. Got back at 5 PM, soaking wet and not sorry to be home.

APRIL 22, TUESDAY. Was a moderately fine day. Employed working out my observations for the positions between Suchartie and this place. A Cutter Sounding and our 2 whale boats absent.

APRIL 23, WEDNESDAY. I had intended trying my walk to the outside Coast again today, but altho there was every promise of fine weather last night and the barometer high, it rained and blew a gale from SE. It is strange these SE winds being so constant at this time of the spring, and the Barometer being no guide. I suspect the remarkably cold winter with const northerly winds have altered the general rule of the weather this year. Mr Moffat, an experienced HB Co officer who has been long on this coast, considers the SE winds to cease early in April and variable winds to obtain till the first week in May, when the NW or summer wind sets in permanently, or did so by his Journal for several years at Rupert & disappearing almost at the same day. These NW winds, he says, die away generally at sunset. In Sept[r] they are again followed by variable winds and mist — the winter gales render the whole west coast a locality to be avoided. Fogs he says may be expected in August but they don't prevail for more than 6 weeks. They roll in from Seaward with the NW wind and may be seen advancing like smoke. They are generally dissipated by noon if there is any wind. We found these fogs ourselves for 2 successive years early in July almost every morning until noon or near it.

APRIL 24, THURSDAY. A very heavy SE gale today which the Barometer did fall a little to ______. We rarely feel these gales much in this Sheltered Anchorage, but the Squalls were so heavy today we let go a 2[d] anchor. At noon it subsided and apparently changed outside to SW. The barometer going up.

APRIL 25, FRIDAY. Moderately fine, but showery in the morning. Walked across to the outside Coast again, and by dint of climbing precipices and Scrambling thro brush, we managed to reach the point from which Woody Cape is seen and one of the offlying Islets to the NW[d] to which I was able to get true bearings and fix my position. There was not so much swell outside as I expected after yesterdays gale, but no place along the Coast where a boat could have landed. At 3 PM we saw our 2[d] whaler returning under sail from San Josef Bay. We reached the Ship at 6.30 PM after a most fatiguing day having taken 4 hours to walk a little over a mile, and 3 to return. It is very difficult getting along the Coast by walking at any time — but unless at low water almost impossible. M[r] Gowlland with 2[d] whaler returned at 7 PM, having completed his work. He says the gale of Thursday did not send such a sea into San Josef Bay but that a ship with good ground tackle would have rode in safety. The bottom is good and decreases regularly from 16 to 5 within a short distance of the head of it, but no shelter is to be had. Sea Otter Cove is a miserable place — with no more than 15 feet of water inside but a passage into it of 5 or 6 fms — the one which was described to us as the Canoe Channel. There appears no place of shelter between San Josef and this place for a boat caught in bad weather unless indeed in the Sandy bay to which I walked across 3 miles NW of the entrance of this harb[r].

APRIL 26, SATURDAY. A very questionable day with Showers. Cutter and 3^{d} whale boat away at work as yesterday. I got obs^{n} for time which enables me to work my observations got yesterday outside.

APRIL 29, TUESDAY. The weather has been anything but settled and SE winds have prevailed. This morning I sent 3^{d} whaler and Cutter with Mr Browning and Blunden to examine the opening called Port Brookes on the Chart, and to regain the ship at the first anchorage found Eastw^{d} of Woody Point. They started early in the morning and I hope got to a place of shelter before the wind sprung up which it did at 8 am.

APRIL 30, WEDNESDAY. Was a misty kind of day, but without much wind, an element the absence of which we earnestly desire at present.

QUATSINO–WOODY POINT–NASPARTI

MAY 1, THURSDAY. I got observations for rates today, intending to leave tomorrow, should Mr Pender not return from Scott I^{ds}, to go in search of him and to run lines of Soundings. If he should return, to steer SE^{d} for the next harb^{r}. At 5 PM he returned and reported having had a severe cruize — strong SE gales had prevailed scarcely without intermission and he had met with some escapes. Had been on Cape Scott for 5 days, weather bound.

The Indians had been very civil to his party. M^{r} P describes a sheltered nook or Covlet on the south side of Cape Scott — excellent for boats with all winds. It is a little west^{d} of the Sandy neck. An Indian village close to it.[127] A heavy surf rolls on the Sandy neck of the Cape with any wind, both on its North and South sides. He had visited Scott and Lanz I^{ds}, found no shelter there and landing very bad. The natives dread the SE gale on this coast. They never venture from C Scott village unless with a NW wind and Ebb (southerly tide) when bound for the I^{ds} 5 or 6 miles dist. There are bad tide rips, near the Cape and Scott I^{ds} both. Mr Pender tells me there is no sheltered anchorage from C Scott to

[127] Possibly Ouchton, a Nakomgilisala village, now IR#3 Ouchton (Galois 1994: 288). The Cape Scott chart shows a village with one house (Great Britain Admiralty 1862).

Quatsino — indeed scarcely a safe stopping place. It is a coast on which the surf almost constantly breaks and off lying reefs extend frequently a mile from the shore. I was glad enough to get our boat back. A canoe accompanied her from C Scott with an old man and woman just married (on their honeymoon cruize). They had 2 lads with them whom they had bought from another tribe for a lot of blankets. These did not consider themselves slaves but kind of adopted Children. They managed the canoe, fished etc. I wanted very much to get the eldest of these lads onboard. He had been of considerable use to Mr Pender in showing him landing places, prognosticating the weather etc., but the old people would not let him go at any price, even for a month. They rarely fail in their prognostication of the weather and study the set of the clouds and general appearance thoroughly before they start on this west coast from one place for another. The SE wind is their great enemy; from the lads just mentioned I learn that the SE season is just over, that we shall get westerly winds for a month, and rain as he says to make the Sting Nettles grow — the natives manufacture lines and canoe gear out of the Sting Nettles, as strong as hemp.

NASPARTI

MAY 2, FRIDAY. At 5 am we left Quatsino and steered to the East^d for Woody pt (Cape Cook) sounding — looked into the opening called Port Brookes on the Chart which appears to afford no shelter for ships. Weather unsettled. Saw a fire on shore which we took for our boat parties. Were abreast Woody Point at 10.30 am, a very heavy swell indeed which I fancy this coast is never free from. A patch of cleared land just west^d of Wood pt some acres in extent which is the only one I remember seeing on the West Coast. It looked extremely rocky. A remarkable Islet about 600 feet high lies nearly a mile off the Cape — it appears double from NW or SE and is I fancy connected with the Cape by a reef. Plenty of whales blowing and leaping out of the water close off the Cape.

On getting East of it, found ESE wind blowing strong, and dirty weather. At one time I thought we should have been obliged to bear up but it went down by noon & we steamed along the land 1½ miles off carrying 20 fathoms least water. When 6 miles East^d of Cape, an opening appeared ahead answering to the description we had had of Nasparti and about

2 PM we entered it. The sea calmed considerably as we got into the bight. This bight or great bay is of considerable depth. The weather was unfavourable to getting a clear view of the coast, but the Eastern side of Nasparti, the place we were steering for, is encumbered by Isld^{s}, rocky I^{ts} and reefs, probably affording shelter within, at any rate to boats.

A remarkable rock 10 or 12 feet high (mile rk) lies off the Entrance of Nasparti due South, mag, nearly 2 miles. We passed East^{d} of it and saw a heavy break to the SW of it. The Entrance appeared free from danger, so we stood in easily — a small wooded, flat and remarkable I^{d} stands in the Entrance. We called it Hat I^{d} and passed a cable East^{d} of it. When within it, the water is quite smooth, the port ran nearly north (mag), and pretty straight, the head of it a Sandy bay is seen from the Entrance. Less than 3 miles off we dropped an anchor in 14 fms, and ran out 6 shackles, dropping a second to moor about ¼ of a mile from the Sandy beach. On sending a boat to sound, we found 10 feet at high water close to our 2^{d} anchor, so weighed both again and steamed out ½ a cable & moored. The Entrance is a little open from this anchorage, but no swell came in, even with a SE gale I think. A considerable Stream empties itself into the head thro the Sandy beach and forms the bank which we nearly ran on. It shoals suddenly from 5 fms to a few feet. Got PM sights on the beach. Several canoes visited the ship.[128] They appear to have had less communication with Europeans than any we have met, and are certainly not very prepossessing people. Most of them had their faces blackened, which may have added to the general effect. The contrast between them and the Quatsinos — not 40 miles distant — struck me as very remarkable and to the disadvantage of these people. They were however not so exhorbitant in their prices and took 2 charges of powder for a Duck or a paddle. However, they will soon take advantage of the great desire shewn by our people for game and who invariably spoil the market wherever they go — in spite of all attempts on my part to regulate a tariff. If they do not do so in the Ship, the absent boats will, so I have given it up now. At Quatsino, nothing short of a Marines Jacket could procure a Goose — and then they would scarcely receive the Marines if they could

[128] Probably the Che:k:tles7et'h' whose territory includes Nasparti Inlet.

get a Sappers — which they much preferred on account of the Scarlet colour. There appear to be numerous Sappers Jackets on this Coast from the British Columbian detachments.[129]

MAY 3–4, SATURDAY & SUNDAY. Weather unfavourable. No sun. Sunday very thick and constant rain. Boats working on Saturday and 2 prepared to leave on Monday morning if fine, one for Esperanza Inlet, the other to Sound outside. This port, which answers to the Native name of Nasparti, is narrow, between 2 & 3 miles long and very straight. High, sharp conical mountains snow capped rise immediately over its head and western sides, and attract all the clouds going which fall in the shape of rain to our disadvantage. Today, Sunday, the Baro is 30.10 — it is strange that SE wind and dirty weather here always followed or accompanied the high glass on this west coast. The SE winds appear late this season. Hoisted the pinnace out and prepared her for service.

MAY 5, MONDAY. Pinnace left to sound outside for 4 days and 2 Whaler to proceed Eastward and examine the Coast for 10. Not a fine morning, and by noon rain and Southerly weather set in. Tried to get something done towards the survey of the port, but was compelled to return.

MAY 6, TUESDAY. A southerly gale and heavy rain. Baro high 30.00. It fell however a tenth yesterday. The SE winds as a rule blow with the Baro at 30.10 & 30.20, tho not a gale. When it falls to 30.00 or thereabout, I remark that a heavy SEly blows. The wind in the harb^{r} is directly south up the reach, but no doubt it is SE outside. Some heavy cascades are rushing down the sides of the steep mountains like mad.

§ NASPARTI–OU-OU-KINISH

MAY 8, THURSDAY. It has rained heavily without the slightest intermission for 4 days, and blown strong from SE. At noon today the wind shifted to North, and at 4 PM the rain ceased. It soon cleared seaward, but the high hills under which we are, keep the clouds and mist still hanging about them. So soon as they cleared we observed that above the height of 1000 feet they were covered with snow. This has taken place

[129] Sappers were the Royal Engineers.

on the near hills within the last 4 days. The distant and higher summits were snow capped previously. At 9 PM the moon & stars shone out and gave me some hope of a fine day tomorrow. I have not seen so many days constant bad weather before on this coast. The sun has not been seen except once for 2 or 3 hours during a whole week. The Pinnace returned this afternoon, having of course done nothing since her departure owing to the weather. We have had no Canoes with Natives since Sunday. The 3d Whaler and Cutter have been detained at Claskino by these late gales, but I hope to see them tomorrow. Otherwise must go in search of them with the Ship. On Saturday morning the northerly wind brought cold — the glass began to rise at noon today. It had not been lower than 29.83 since 9 PM 30.00.

MAY 9, FRIDAY. At last something approaching to a fine day. I got equal acts latitude etc. and in the afternoon the Cutter and 3d whaler returned from Claskino. They had experienced SE gales and constant rain like ourselves for 6 days and had done but little; but Mr Browning reports the place much more extensive than we anticipated and having found an anchorage for the Ship, there I must move round as soon as we get a little more advanced with the work in this direction.

MAY 10, SATURDAY. It blew heavily from NW yesterday and today with clear weather — so heavily that the boats found difficulty in moving about and it was almost impossible to use a theodolite. I went to the harbr entrance, visited a small village[130] near the eastern point of the harbr, but could not get outside far for the wind. 1t Whaler went [to] the western side, but was not able to go far out.

There don't appear to be many natives here, I think not over 50 altogether. I observed a Spanish cast of features about many of them. They have a little fish and some bad Ducks to dispose of, and appear ill clothed as far as blankets etc. Most of them were in the bark mats made of the Yellow Cypress. Truly this is a wonderful wood. They make paddles, and all their articles of furniture of it. It completely clothes them, and I think would make good rope. I got some of the twine made of it, also some of

[130] This is likely the village of Acous, now a reservation of 40.5 hecatres (Acous IR#1).

the fibre of the nettle — the latter much the strongest, I should say as strong as flax quite, if not stronger. The canoes are made out of the Cedar, which grows to a larger size and is far more plentiful than the Cypress. I got a little of the latter here, but had not much time to seek for it. The language of the Indians here is something similar to the Barclay Sound dialect, and M[r] Hankin is able to make himself understood, knowing the latter.

It is somewhat singular that the 10th of May was the first day of summer — last year as it is this. Winter has taken its lease with most abominable weather. SE gales, rain for 5 consecutive days, and then summer is ushered in with gales from NW, so strong as to be almost as fatal to our work as the positive bad weather was.

MAY 14, WEDNESDAY. I have been away in my boat since Monday morning, examining the Coast Eastw[d]. Found it cut up with Inlets and arms in a most singular way, entailing at least a months more work than I had bargained for. For 30 miles SSE of this place the coast is studded with Islands and off lying reefs without number, the latter extending some 3 miles off the land. Those near seem so remarkable a cluster of dangers. There are 4 or 5 deep inlets penetrating the Island within the same distance, that is, between Nasparti and Esperanza Inlet, some of them 20 to 30 miles in length — many sunken dangers lie off the islands and it will require great care and labour to survey this space. I went some miles up the first opening East[d] of this "oo-oo-Kinsh" [Ououkinsh Inlet] — but finding no end to it I employed myself working outside, where it was only possible to land during intervals of fine weather. I have never seen a Coast so cut up as this is — with such an infinity of rocks lying off it, most of them from 2 to 10 feet out of water, many sunken.

M[r] Gowlland returned this evening in 2[d] Whaler, and his report and Sketch of the Coast almost bewilders me — and obliges me to considerably modify my plans. I am going to return west[d] to Claskino or Port Brooks in the Ship on Friday, leaving 2 boats to work eastw[d] on the Coast and to meet me at Esperanza Inlet in 10 days. The great inlets between this place and that must for the present remain unsurveyed. M[r] Gowlland has found a safe anchorage in Esperanza for the Ship.

Bought a Deer today for a blanket, a high price but as we are in rather short commons and Venison Scarce, the opportunity was not to be missed.

The NW winds have blown with great violence since they commenced on the 9th, so much so as to prevent our Sounding boats from Working frequently. They have not died away until very late at night and I am doubtful if they have not blown with their full violence a few miles off shore. I suspect even close to Woody pt. The weather has been very clear with them.

WOODY POINT

MAY 16, FRIDAY. At 4 am we left Nasparti. The 2d Whaler and Pinnace with Messrs Gowlland and Blunden starting at the same time to the Eastward. It was a very fine morning and we had a good opportunity of fixing the positions of the mountains on the Promontory[131] which separates this harbour from Klaskino, the extremity of which is Woody Pt off Cook, or as I call it, Cape Cook. A very remarkable double peaked islet[132] 500 feet high about lies off the Cape, connected with it by reefs; rounded it at 8.30 am and steered in for Claskino Cove. Mr Browning, who had examined it previously, acting as pilot. The cove or anchorage is formed by a remarkable Nob — wooded point on the East and a still more remarkable mount[133] on the west, down the wooded side of which is a land slip from summit to base shewing as a red stripe 4 or 5 miles to seaward. It must not be confounded with another Mountain 2 or 3 miles westd which has also a somewhat similar slip bare of trees down its side — but it does not present the same red colour nor is nearly so remarkable.

Our track in lay between some high rocks extending off and connected with the Tree Nob point before mentioned and a reef a few feet above water westd of them. The Channel is ½ a mile wide and looked quite deep, so that I felt no hesitation in steaming in, tho it had not been Sounded. Mr Browning had seen it in heavy weather and observed no

[131] Brooks Peninsula.

[132] Probably Solander Island.

[133] Red Stripe Mountain.

break. As we were in the middle of the passage we observed a swell close ahead of us and the ship rolled very heavily. Stopped her immediately — struck 6 fathoms in front chains — the same over the stern with the deep sea lead, but no bottom; the starb[d] side in 20 fathoms. As we passed inside the reefs the swell ceased and we soon entered the cove by a very contracted Channel, which required some care in threading but still was quite wide enough for a Ship of our size; and when known one would have felt no hesitation in taking a ship of any size thro it. Moored in 10 fathoms, with plenty of room 1 cable from the nearest shore, room for 3 ships of our size moored.

⸙ KLASKINO–KLAS-KISH

MAY 17, SATURDAY. Sent a boat to sound the Entrance between the 20 foot Rock and west reef by which we entered yesterday. Found it strewed with rocks, some of which break near low water. We must have had a most narrow escape in entering, so much so that it will not be an easy thing to take the ship out by the way she came in after the Channel has been thoroughly examined. It is perfectly wonderful how we escaped. Got observations on a Sandy beach where there is the remains of a large Native Village.[134] All boats away examining and Sounding the neighbourhood. There appears to be another anchorage 4 or 5 miles east[d] of this, called by the Natives Klas Kish [Klaskish Inlet]. But this whole indentation is not to be recommended, tho doubtless it affords anchorage and refuge to vessels damaged or on a lee Shore, should they have a chart of it. Otherwise, it would be almost certain destruction to run for it.

Some Natives gathered here from Klaskish, attracted by our presence — and a large canoe with some of our Quatsino friends met us off the Entrance and came in, pitching their tents at the large Native Village or rather Skeleton of a Village. These large massive Lodges would appear to be the permanent Winter residences. The Natives leave them in summer and wander about wherever fish and berries are most abundant, merely throwing up temporary residences of mats or sails of their canoes. 10th May may be considered the first day of summer on this Coast — it

[134] Richards is in Klaskino Inlet. According to Chart D6841, the houses are located on the eastern entrance to the inlet. This is currently recorded as IR#2 Tsowenachs.

has been so for 2 years and the Natives know almost to a day when summer has really begun and the SE gales and rain over. They never begin to move about until about that day. I am not certain whether I am right in supposing these large massively built Villages as their permanent winter habitations or whether they are the abodes of tribes extinct & totally abandoned. They appear much worn, yet the Natives now who have come here during our stay make use of the Village — adding their own mats and gear to make them habitable.

Got stars and other observations here. Weather very fine and clear. Some remarkable cone shaped[135] hills rising immediately over the shore. This is the general feature between Quatsino and Nootka. Evidently there has been some great convulsion within these limits. It is a perfect nest of reefs between the Coast and 3 miles seaward and the conical shape of the Mountains is no less remarkable land ward.

MAY 18, SUNDAY. Very fine day. All boats prepared to leave early tomorrow morning. At 8 PM a bank of clouds heavy to the west[d] and a perceptible rolling motion of the Ship in this small cove made me doubt what tomorrow may bring forth.

MAY 19, MONDAY. All boats away early. 1 Whaler went to Clash-Kish — Cutter Sounding, 3[d] Whaler looking for the reefs in the channel. I went myself also to examine the Entrance. We found the two breaking reefs right in the centre of the Channel and it is still more extraordinary to me how we escaped them. There are some others yet to be discovered.

MAY 20. Sent cutter to join 1[t] Whaler at Klash-Kish. Weather a little gloomy today, threatening a change; it blew very fresh from west[d] for 2 hours, then subsided and cleared again in the evening. Got obs[n] on shore and measured a base. A party went to the head of the arm and remained away 24 hours, hoping to get some game but tho they saw numerous fresh traces of Deer, they got nothing. A party for Shooting Deer should be small. No tent with them, no fire, no Cigars, but build a wigwam of branches and lie quiet, taking a position where the animals are likely to

---

[135] Richards pencils in an almost illegible footnote here which seems to compare the conical shape of the mountains with "the shape of the young people's heads."

come down for water. The natives have gone away today. I fancy they only came here with the object of making what they could from us — which is not much, the ship is almost deserted so many parties being away; and the natives have little to dispose of. There appears to be three dialects on this West Coast — at Quatsino they speak the Rupert language. Here and between Woody Point and Nootka inclusive it is something like the Barclay Sound. Our great chance is to find a native who can speak the Chinook Jargon — we have one onboard now who can talk it a little. He is an intelligent kind of lad, belongs to Kayuquot about 25 miles east^d^ of Woody P^t^. He dwells much on the grandeur of his Chief who he says has any quantity of whisky. The west coast was almost free from this Curse until Europeans began to frequent it, which they did as soon as the HBCos exclusive right to trade in Skins ceased and even before. Now small vessels frequent all the native places and whisky I believe is their principal article of exchange. The Surprise is I believe the keenest trader on the Coast commanded by a well known Cooper named McKay. He gets oil and Skins from the natives and is much liked by them. I believe he treats them always well and fairly — and finds it the best policy. As regards his providing them with whisky, I imagine it is no more than can be expected from a man of his pursuits; the oil and Skin trade is very lucrative, and we must consider the inducement before we condemn a man for the means he adopts to make his fortune. It would seem rather the duty of the authorities to interpose some check, either by placing proper officials at each of the harbours, or perhaps better, by establishing native missions. The whole of the west coast of the I^d^ is allowed to take care of itself, and it is not difficult to see which way their morals are tending — as their communications with ourselves increase. Kayoquot is said to be the most populous district on the West Coast. 600–700 people there (I cant make out above 8000 on the whole I^d^) and our Indian who belongs to the tribe says that whisky is the only thing they will sell the skins or oil for. I dont quite take all this for granted, because I have always found that the great desire has been for blankets.

It is somewhat remarkable that during five years we have been on this Coast we have scarcely met with an instance of dishonesty among the Natives; they have had every opportunity of pilfering from us, and yet the instances have been extremely rare. I only remember one decided case, and this at Nootka Sound where 2 Indians ran away with a Chronometer from me while observing on shore. I had left it a few yards from me covered over with a cloth while I observed at another Spot for Magnetic data. The watch was recovered and it turned out that the 2 men who took it were foreign to the tribe. The act was thoroughly disavowed by the Chief Maquinna — and gave them much annoyance.

The only ones among them who seem to have the slightest idea of dishonesty are those who have visited Victoria and become acquainted with European customs. At the same time it is only fair to say that it is scarcely right to judge by what happens to us. They naturally dread our force, and may be deterred from taking advantage from this cause. Still, our boats are constantly among them far removed from the ship and quite in their power. They have been always foremost to help them, landing in a Surf a whole village has come down to haul the boat up or to launch her, and they have always shewn the most friendly feeling. I can safely say, having seen & had dealings with almost all Native tribes in the world. I have never met a more friendly, harmless and well disposed set of people than those on Vancouver Island.

KLASKINO–QUEENS COVE, ESPERANZA INLET

MAY 23, FRIDAY. During our stay in this place the weather has been remarkably fine — indeed since the 10th of May there has not been any rain, tho on 2 or 3 occasions too much wind to work outside from N.W. The first burst of these winds would appear to be the strongest. The natives returned and set up to the number of about 20 in the village close to us. They bring great quantities of Hallibut — some weighing 150 pounds. They are a coarse fish, but the men almost live on them. There is little else to be had. Yesterday the 2 boats returned from Klash kish in the SE corner of the bight formed between Woody Pt and Quatsino. They found it a snug little place, better than Klaskino and more free from danger in entering. There appeared to be a change of weather last night and it rained a good deal. On Friday morning at 4 am however we moved out of the cove and steamed out of the passage, passing about a cables length westd of the 20 feet rock off Nob pt. A SW½W course there leads out eastward of the sunken rocks. The weather clouded over when a few miles outside with drizzling rain and the land obscured.

Wind SEly. I thought we should have had to bear up for Quatsino at one time, but as the wind did not freshen I had an idea that the gloom and mist would be dispelled as the sun got higher — and was not wrong. At noon it cleared and the afternoon was beautiful. I have never seen so little swell on the west coast before — we had been getting deep Soundings all day. As I before judged, the great bank which extends so far off Vancouver I[d] does not commence until South[d] & East of Woody point for 7 or 8 miles. At 6 we were off the Entrance of Esperanza Inlet and, with the assistance of M[r] Gowllands sketches, we steamed in and found our way into a very snug cove on the east side of an arm marked on the chart as Port Eliza — the Entrance to the cove is very narrow, about 80 yards, but within it is wider and we moored in 10 fathoms. I called it Queens Cove, as Her Majestys birthday occurred on Saturday, the day after we entered. Great numbers of natives visited us from the neighbourhood of Nuchatlat [Nuchatlitz Inlet] and E-hasset [Esperanza Inlet], the former an inlet on the NW side of Nootka I[d]; the latter belonging close here. The father of Macquilla, Chief of Nootka, was among them.

MAY 25, SUNDAY. M[r] Hankin and D[r] Wood went to Kayuquot by canoe, with the view to crossing the I[d] to Nimpkish River where I purpose meeting them with the Ship on the 16 June.

JUNE 1, SUNDAY. During the last week the few left onboard have been at work at the Entrance of this Inlet. For 3 days of the week we had SE gales and constant rain. The 2[d] Whaler and Pinnace returned on Saturday, having completed the outer coast between Nasparti and this place. M[r] Gowlland tells me he met M[r] Hankin and Doctor Wood in difficulties. It appears that the Ehasset natives who took them from here with the understanding that they were to convey them to Kyuquot, landed them about 3 miles from this and then left them, and laughed at them, saying they were quite far enough. They had been paid 5 blankets and other things for the service. The D[r] & M[r] Hankin, not liking to return, managed with difficulty to get a very small Canoe with 2 men to carry them and their traps to Kyuquot, and were fortunately met by M[r] Gowlland on the way and taken by him to the Kyuquot Village among the I[ds]. And fortunate it was further, for the canoe was so small and the weather so doubtful that in all probability they would have reached nowhere but for this fortunate meeting. M[r] Gowlland from his acquaintance with the Kyuquot people was able to help them considerably and I hope has ensured their reaching Nimpkish. I made a point of finding the natives who had behaved so knavishly and obliged the Chief of the Ehassets to return the blankets paid by M[r] Hankin, who protested himself much annoyed at the behaviour of his people. After getting the blankets back and making them fully understand my disapproval of their dishonest proceeding, I made the Chief a present of the blankets so received, or part of them. I have rarely known an instance of this nature. M[r] Hankin was wrong to pay them beforehand, and all arrangements of this kind should be made thro the Chief. M[r] Gowlland describes the Kyouquot tribe as very numerous and friendly. There are 400 fighting men and he says 1600?? in all. Many more women and children were observed among them than in most other parts. The story of our Chronometer having been stolen at Nootka 2 years ago was well known among them. Indeed, the fame of that transaction seems to have spread all along the Coast, probably from the fact of my having kept the 2 delinquents in Irons for a week more than from the horror of the transaction in a moral point of view. However, the Chief at Kyuquot warned his people to steal nothing from M[r] Gowllands party — nor to lose their good name as the Nootkas had.

My old friend Macquinna came here from Nootka to see me and brought me a Deer as a present, which of course involved one of a more Valuable Character, and I accordingly gave him a blanket and some other things. I did not know him at first — he was dressed in a blue frock coat with 3 rows of buttons — one row American Eagles the other Royal Marines. A pair of black cloth trousers and <u>over all</u> a long black beaver hat or <u>Bell topper</u>. He was so totally disguised that I did not know him in the least for 2 days. Moreover, as he was accompanied by a wife totally dissimilar from the lady who filled that office in 1860 I had no suspicion in the world that it was him, tho when he took his hat off I remarked a likeness to Macquinna. He was evidently I found afterwards hurt at my not recognizing him — and after 2 days came onboard with his common dirty blanket on — then I immediately thought it must be my [illegible] friend, but on enquiry from his wife who spoke a little Chinook, I understood it was the brother of Maquinna. I felt even then that

I had been very wanting in attention to the family and tried to make amends for it by supplying them with food. My time, however, was so entirely occupied at this place with the Survey that I could not pay much attention to social duties. On Mr Gowlland coming back yesterday, he and Macquinna immediately embraced each other. M^{r} G had met him some few months since and his memory served him better than mine. I then made all amends in my power. Macquinna was perfectly satisfied — and he & his wife & 2 dependents who had brought him from Nootka were installed onb^{d} and provided with food. I also offered to take them back to Nootka where I am going in 3 or 4 days.

JUNE 2, MONDAY. Today I finished all I could do towards the Survey of this place and tomorrow leave for Friendly Cove. There are some beautiful spars on a piece of flat ground at the head of this Cove. We got 2 onb^{d} and cut 3 more in case we come here again.

The number of natives in this inlet as well as I can ascertain among the three tribes E-hash-ets, Nu-chat-lets and Ao-quas are as follows — E-hash-ats 70 fighting men — 110 women & Children; Nu-chat-lets 60 fighting men — 95 women and Children; Ao-quas 82 fighting men — 120 women and children. These numbers are, I believe, very correct. The natives enumerated them by a piece of stick for each individual, and named every man with his wife and family as he counted them. Perhaps the Ao-quas tribe are less correct than the other two as regards women and children — which latter they only estimated but I suspect they are very nearly correct.

I observed that Maquinna had a different wife to the one he possessed on our last visit. I questioned the present lady as to the reason. She told me he had two — and the 1^{t} was at Nootka near her confinement (Chiefs are allowed the privilege of a plurality). I was rather surprised at first that he should have become enamoured of the present lady. She was by no means young, nor good looking and moreover had been the wife of his father still living. I imagine it is an alliance of convenience. She speaks the language of the Eastern Coast and a little Chinook and is therefore useful to him in his dealings with the schooners which come for oil etc. She informed us that Maquinna had given 10 Sea Otter Skins for her to the Rupert tribe; his version was one Skin, probably nearer the truth and more than she was worth. They came onboard this evening and we hoisted their canoe in.

1862 ESPERANZA INLET–NOOTKA SOUND

JUNE 3, TUESDAY MORNING. We left Queens Cove, Pinnace being detached at the same time for Kyouquot. She first towed our head round for us as the place was too narrow to turn in. Steamed out thro the Centre Channel — not the one we entered by. 10 fathoms the least depth, but a tide rip and confused sea and not a very pleasant passage until well Sounded, which it was not. However, we had a deep channel with a leading mark; Black Rk Φ [between] Round Hill 800 feet high bearing NthW, and steamed out a SthE course. Met strong SE wind with thick dirty weather outside, and were obliged to work on the first step to make headway. Carried a line of Soundings down about 3 miles offshore. Saw the Bajo Rock breaking, indeed uncovered, about a mile off the Shore and were passing about 1½ miles outside it when we suddenly shoaled from 26 to 16 fms and then to 10. Kept out 4 points and shoaled to 6 fathoms. Kept straight off the land got 8–9–10 and a mile of 20 fms. Just as we kept out saw 2 breakers about ¾ of a mile from us, right on the course we were steering in the first instance when we shoaled the water. Fixed them and found that they lay 2½ miles S28E Mag, from the Bajo Rock above water. We had passed down on a former occasion within a cable and a half of these breakers, but being smooth water had not seen them and but for the heavy sea today should not have judged that any shoal water existed outside the bare rk at low tide. Vancouver says in his work that the bajo extends 2 miles off shore — it is 3 miles off, and vessels passing along this coast should give it a berth here of 4 miles. There is 14 fathoms between the inner uncovering rock and our breakers of today. Mr Browning passed between them in his boat and carried this depth but vessels should by no means pass between them.

At noon we bore up for the Entrance of Nootka Sound. Got a good View of Conuma Peak bearing N32E Mag, although the weather was so gloomy and unfavourable. Ran into Friendly Cove and moored in the centre of it. There is plenty of room for a vessel of this size moored to right — and right amidships. Drop the 1^{t} anchor in 9 fms — when abreast the Centre of the outer I^{d}, Sulphur I^{d}, & steer in for the middle of the Sandy beach. Run out 7 shackles and drop the inner anchor in 4½. Landed and visited the village, which assumes a very different aspect to what it did on our last visit in winter. Then, the bare poles of the village alone stood and we should have considered it deserted; now it was all boarded in and contained very commodious houses and cleaner than any village I have ever seen. Maquinna and his old lady landed also, and we visited them. I insisted on seeing the former wife and after much difficulty she appeared, looking very clean and nice. Evidently there was some estrangement between her and Maquinna. I must endeavour to ascertain the cause and if possible effect a reconciliation — for it is very unsatisfactory for me to see her supplanted by the ancient lady who appears now to have an unbounded sway and control over him. I think the seal, for so we call the young wife from her likeness to that animal, was struck with the interest we shewed in her, and somewhat surprised at being remembered at all.

The uncle of Maquinna — and relative of old Maquinna who was Chief when I was here in Sulphur in 1839,[136] came onboard today with the other chiefs. He remembered the visit of Sulphur and Starling, and when asked about them immediately said — "Belcher", the name of the captain of the expedition. This was strange their remembering his name after so long an absence, and we were only here 4 or 5 days in Sulphur. They asked my name, but had some difficulty in pronouncing it and kept repeating it over several times. With the ships they were more successful — and could say Hecate remarkably well. They brought me onboard a Sea Otter Skin as a present, and on the occasion observed a Ceremony I have not before known. 8 of their Chiefs stood in a line on deck when one asked my name, then after a loud oration presented me with the Skin as a "potlatch from Maquinna". He, Maquinna, did not take part in the Ceremony, but stood by in the Coat with <u>3 rows of buttons</u>. I gave them some biscuit, rum & water, and pipes and tobacco, and told them to come tomorrow when I suppose I must make a present in return. I am bound to say the Otter Skin is the smallest I have seen and not

[136] Richards visited Friendly Cove in 1837. In 1839, the *Sulphur* was once again on the coast, but Belcher's narrative does not indicate that they stopped at Friendly Cove during their ten-day trip from Sitka to the Columbia (Belcher 2005: 287–288).

of any great value — but the gift involves a return of double its worth, besides the waste of a good deal more time in shewing them attention than I can afford.

I have been able to satisfy myself during the last day or two of the relationship existing between the present reigning family and the Maquinna of 1839 without any doubt. The man I call Uncle of Maquinna (the present young chief) is a son of old Maquinna — the same lad we saw in Sulphur & hence his perfect recollection of all transactions at that time. He was then 18 or 19 years old. The young Chief is the son of the girl mentioned in Cap Belchers — Voyage — the daughter of old Maquinna. Why the succession should have gone to her son before the son of Maquinna the elder, I dont know, except that I have frequently found the chieftainship descends in the female line, probably when she is of greater rank than the male. At any rate, the Uncle of the present lad appears to me to be far more fitted by his character to be chief — than his nephew. He is sober, dignified, and evidently a man of strength, while young Maquinna is neither. He is fond of rum, fine dress etc, and I think has no great influence over the majority of the tribe. I fancy a Chief holds his authority in virtue of his property — or the number of blankets he possesses. If he did not succeed in accumulating blankets, he would soon sink into insignificance. He retains his influence over the people by the presents he makes them, hence we find Chiefs giving away hundreds of blankets at a time. When a man can accumulate enough blankets to fill a half a dozen large boxes, he is pretty certain to be acknowledged as a head, as in the instance of our friend Jim of Fort Rupert. A Chief must be very prodigal — be able to tear up 2 or 3 Sea Otter Skins, and give away the pieces, or throw his whole wardrobe into the public treasury.[137]

I have been taking great pains to see the girl who we met in Sulphurs voyage here — and I find now that I did see her at Esperanza — and shewed her no attention whatever. I remember noticing her perfectly as

---

[137] It was a practice among many First Nations on the Northwest Coast to destroy property as a display of wealth. According to Drucker (1963: 137) "the destruction of property . . . was to demonstrate that the chief was so powerful and so rich that the blankets or money he threw on the fire . . .. were of no moment at all to him."

a tall, stately looking dame very much whiter than the generality of the Natives — but at the time was so engaged about other matters that I had not time to devote attention to her — indeed, never suspected who she was, tho I knew very well she was the wife of the man who told me he was Young Maquinnas father. This stately lady is his mother and the little pretty girl we saw in the Sulphur. I shall yet I hope have an opportunity of showing her some attention — but their language is so difficult to get hold of, and the few who can speak Chinook among them make it a puzzling affair to trace any connecting link — particularly when our time is so wholly engrossed in other matters. I ascertained from Macquinna, the Young Chief of Nootka, that the numbers of their tribe (including all in the Sound) is as follows — 199 men — 203 women and 57 children, 459 in all. They gave me 3 bundles of sticks according to the Native custom, one bundle of men, one of women and one of children.

KYUKUOT SOUND

JUNE 8, SUNDAY. Was our last day at Nootka. The Natives literally lived on our decks all the time we remained, and were very loth to leave at the last. We gave them every thing we had in the way of presents and they made a good harvest by us. I promised to return in a month, which I hope to do.

Sunday night was very fine and Monday morning 9th was equally so. Fresh land wind from NE, but barometer had gone down 2 tenths during the night and at 4 am stood at 29.78. I had some doubt about weighing, but decided to go — so we unmoored and left soon after 4 am. At 8 a SE wind sprung up and increased to a gale. We carried a very useful line of Soundings about 10 miles offshore from 50 to 70 fathoms. The sea and wind and thick weather increased so much at 11 am that I [illegible] the storm sails & made up my mind to stand off and bank fires, particularly as the Entrance to Kyuquot lay between a long line of reefs on either side and had not been examined. I was induced, however, by the prospect of a snug anchorage, to stand in to the outer extreme of the reefs at the entrance, and the channel looking clear enough. We went in. The gale increased very much as we entered and rain with very thick weather came on so much so that ¼ of an hour later and we could not have entered. Met the pinnace inside — she had preceded us to this place a week before to examine the channel, but the weather was too bad for her to come outside to us; at 2 PM we anchored in a narrow creek on the eastern side of the Sound about 6 miles within the entrance and moored, the place being too narrow to lie at single anchor. It rained all day and the clouds were passing rapidly from SE. This is a remarkable little anchorage not above 250 yards wide. Mountains rising to 3000 feet immediately on either side of it. The depth of water 12 to 16 fathoms but very shoal at the head where 2 rivers flow into it. No Natives here. There are more numerous in this district than at any other part of the West Coast — but are now living among the I[ds] and I suppose in the thick weather did not see us come in yesterday.

JUNE 10, TUESDAY. Prepared the 3 boats today to leave behind during our return to Esquimalt and provisioned them for 2 months. Fired the pivot gun to bring the Indians up. 2 small canoes came but they could not have heard our gun at their islands 12 miles from here and the weather was too boisterous for the small canoes to go off to them.

Some good streams running into this harb[r]. Watered from one of them 100 yards from our anchorage. Unmoored in the evening and made the Ship fast to the trees to steady her, ready to start at 4 in the morning. I got observations here at the head of the Creek. Barometer up to 30.20 this evening. Rain falling but I suspect only local, surrounded as we are by mountains. It is very remarkable the prevalence of SE winds at this season when nothing but NW[ds] were looked for. We have had 3 days of SE gales and rain regularly in a week. The natives look upon these winds as their natural enemy. They call it Mitlass, and always speak of it as pe-shak, which means very bad. The NW wind they call Tsa-qua, it always brings fine weather.

1862 KYUQUOT SOUND–SCOTT ISLANDS

JUNE 11, WEDNESDAY. At 4 we left — it was low water and near the springs and the place was so narrow that there was only just room for us to tow her head round with her stern made fast to a tree on the opposite side. Dropped the 2 whalers and towed the pinnace to the entrance where we left her. Steamed out of the channel and to the west[d] carrying a line of soundings, and getting lines to the mountains which shewed beautifully clear. I have rarely seen a more beautiful morning, and as rarely seen a fine day throughout in this country — this one was not doomed to be an exception. As soon as we rounded Woody point at 11 am a fog bank was seen to the NW. We passed rapidly thro 2 or 3 narrow belts of it and then into the thick of it. All land was immediately obscured and a chilly damp NW wind succeeded the bright sunny morning. This was the more provoking as we were bound for the Scott I^{ds} where we should have to knock about all night — and this was our third attempt to complete the survey of them. It thickened towards eve — we steamed up and made the Inner I^{ds} between Scott and Lanz. Saw a point of land for a few minutes, just long enough to fix our position and the line of Soundings. It was about a mile from us. Then we steered South (true) under very easy steam till Mid^{t}, when we went NW again and so on till daylight when the fog remained as thick as ever.

SCOTT ISLANDS

JUNE 12. Very thick fog all forenoon. Sun shewed at noon and enough horizon to get a latitude. Saw the Triangle I^{d} also at noon, and found we had been set some 12 miles to the west since last night. The flood tide sweeping along the shore west^{d} and the ebb out of Goletas Channel both contributed to this. Kept sounding constantly in depths varying from 240 fathoms to 80.

It appeared likely to clear after noon and we ran along the South side of Triangle I^{d} flattered with the hope that we were going to get it into its position on the chart with its off lying rocks — but were mistaken. It came on thicker than ever, and we soon had to look to our own security rather than providing for that of others who are to come after us. A tide race appears to extend off the SW end of the Triangle I^{d} for 2 miles — this is the full moon and consequently spring tides, the most critical time for knocking about among unknown dangers in a fog. Soon however all was entirely obscured and we steered in towards the land of VI about Sea Otter Bay, feeling our way with the lead. These fogs are a great drawback to the navigation of this coast and the entrance of Goletas Channel, and I believe are much more prevalent than generally supposed. They are said to prevail in August & Sep^{t} — but I have found them here early in July, and now in June. In short, I have never come to the neighbourhood without meeting them and this is the 3^{d} time I have been baulked in the examination of these I^{ds} in consequence of fogs.

The prevalence of these great drawbacks and dangers to navigation should be taken into account before establishing a port at the North Bentinck arm.[138] It is inconvenient enough entering San Francisco, with a comparatively quiet climate and coast pretty free from danger but even here winds are common.

On a Coast like this Strewed with dangers and subject to strong Currents and furious gales with thick weather, it is certain that many serious and fatal disasters must be looked for — should sailing vessels or even steamers engage in a line to the Bentinck Arm or any part of the Coast West^{d} of Vancouver I^{d}.

To steamers passing up inside VC I^{d} and starting from Fort Rupert under favourable circumstances, with a knowledge of the Coast, it might be difficult, but as certain as sailing vessels attempt the route from San Francisco to these Sounds in the neighbourhood of Bentinck Arm, so certain will there be many fatal accidents.

1862 SCOTT ISLANDS–GOLETAS CHANNEL–FORT RUPERT

Banked the fires and put the head to SW for the night and at 4 am of the 13th got steam and steered NE to make the land. At 7 am made Sea Otter I^{d} uncomfortably close thro the fog and stood off to the west^{d} Sounding. At 9.30, having crossed and conducted a former line of Soundings in 100 fms, steered for Scott passage which we soon made. The weather continued thick but the fog not so dense as it had been. The Barometer very low and falling, 29.97, and being short of coals and provisions obliged me to abandon the examination of the Triangle I^{d} for the present, and return to Fort Rupert. Found the flood (about ½ tide) running to the

[138] In 1873 North Bentinck Arm, although an unlikely candidate, was one of several harbours under consideration as the terminus of the Canadian Pacific Railway (Berton 1970: 194).

NE thro Scott Passage about 1½ miles an hour, and when in the falling straight thro the Channel of Goletas at same rate. We were on Newhitti Bar at H water, consequently did not feel the full strength of the tide which here is 4 to 5 knots at Springs. Ran thro Goletas Channel, a strong NW wind following us, and entered Beaver Harb^{r} by the western channel inside the I^{ds}. Anch^{d} at 4.30 PM. Found a calico house erected over my OB on Shell I^{d}, which I at first took for a whisky sellers establishment, but soon found it was the grave of a young Tyhee girl of the tribe who had died at Victoria and been brought back here. Lt Hankin and Dr Wood rejoined us having succeeded in crossing the I^{d} from Kyuquot to Nimpkish after many difficulties, both with the Indians and from floods and rapids. They were 8 days accomplishing the journey — the Indians most unwilling to proceed, their provisions had run out, and the D^{r} was much knocked up. The season for travelling was too early.[139] We found the small pox raging here among the natives, who were much subdued and terrified by it. 16 cases had occurred up to today, 5 of them proved fatal. As soon as they are attacked their friends leave them to die frequently putting them out into the bush, and scarcely giving them burial. Hundreds presented themselves aboard and eagerly demanded to be vaccinated.[140] Our Doctors did all who applied. Many had been only done a few days before by Mr Moffat of HB Co, but were desirous of being re-operated on — which of course was not done. They had no

[139] Hankin's account of this journey is in the BC Archives Colonial Correspondence BO1349, 1215 Navy-H.M.S. Surveying Ship *Hecate*. It was also published in its entirety in the *British Colonist*, December 13, 1862.

[140] A smallpox vaccine was developed in the late eighteenth century from cowpox, a related disease less lethal to humans. By 1836, the vaccine was being administered by Russian doctors to First Nations on the North Coast (Boyd 1999: 121) and by 1837, the Hudson's Bay Company was administering it on the Northwest Coast (Boyd 1999: 134). However, vaccination occurred primarily around trading posts and forts, and did not reach more remote First Nations. It was also often difficult to convince First Nations people to be vaccinated. The epidemic that Richards witnessed had devastating effects. According to Boyd, nearly 14,000 Aboriginal people died in the 1862–63 smallpox epidemic on the coast (Boyd 1999: 172). Although a vaccine was available, it was in short supply and other methods of control, like quarantine, were not employed when the epidemic began in Victoria. Instead, infected people were forced to leave, spreading the illness to others along the coast.

idea of any kind of gratitude for the service rendered them, and in some cases wished to be paid for going thro the operation. The village is deserted — they have distributed themselves about in 20s and 30s, apparently afraid of each other. All fishing or hunting is at an end, and Mr Moffat tells me that he thinks 3 or 4 have died from actual fear, for no traces of the disease could be seen on them — and they died on the 3^{d} day after being attacked.

JUNE 15, SUNDAY. Three fresh cases of small pox occurred on shore today. One poor girl being seized with it, was immediately removed to a distance from the village and placed under a kind of hut or wigwam built of branches, and there left. It is distressing to be able to afford them no assistance or relief — it could only be done by erecting a building to receive them and supplying medical assistance and comforts, neither of which are in our power, but govt ought to do something, funds or no funds. We leave at daylight tomorrow morning, and even if we remained have no way of alleviating their sufferings.

The stern wheel steamer Flying Dutchman with 50 passengers arrived this evening from Victoria and Fraser River. Got some papers by her. She is bound for the Stikeen River diggings.

1862 FORT RUPERT–NANAIMO

JUNE 16, MONDAY. 4 am left Rupert and steamed for Johnstone Strait. Passed 3 boats full of passengers for the Northern mines I suppose. A strong Easterly wind sprung up at 8 am against us. It slackened during the day and rained a good deal. It was just after hw when we left Rupert (5 days after full moon) so we have the whole ebb against us and as the tides are 4 hours later at C Mudge than Rupert, we enjoyed nearly two ebbs against us.

At noon we were abreast Port Neville. Ebb strong abreast the Salmon river. 4.30 PM at Seymour Narrows. Strong flood with us and much race. Passed the middle of it, quick and careful steerage necessary. About 3 PM would have been slack water today. At 5.15 PM abreast C Mudge. The large village just inside is now built up and inhabited. I counted 50 canoes on the beach. This is one of the rare instances of a Stockaded village.

JUNE 17, TUESDAY. 3.30 am moored in Nanaimo. Found all the Natives had been removed from the Town for fear of small pox.[141] The Forward had started 2 days before to convoy a troop of Northern Canoes to Rupert. She must have been at Komox when we passed. We coaled at Nanaimo and the weather was fine during our stay.

§ NANAIMO–ESQUIMALT–NOOTKA SOUND

JUNE 22, SUNDAY. At 6 am we started for Esquimalt. Not a pleasant day, rain & SE wind. At 7 PM anchored at Esquimalt. Found the Admiralty in Bacchante, and the Charybdis here. We remained here until the 8th July. The weather very fine and warm. Two mail steamers arrived during our stay, and two emigrant vessels from Otago, N Zealand, allured here by the reports of the Cariboo gold, deluded, I should say. I am fully occupied here in Courts Martial, Courts of Enquiry, attending on the admiral, <u>and in doing nothing</u>. Certainly I had not a moment to myself the whole time — scarcely recd my letters and scarcely answered any. Glad enough to leave.

> JUNE 30. Started at 8am from Camp; Pinnace taking a line of soundings down the Inlet; and then to rendezvous at the Village of Actsae[142] on the sea board preparatory to taking in her remaining provisions left a Month ago in Charge of the Chief. The 2nd Whaler made a plan of the Anchorage on a 6 in scale on the west shore of this Arm of the Sound and Blunden to make a plan of another Anchorage discovered by [illegible] on Centre

---

[141] John Booth Good, the Anglican missionary at Nanaimo, wrote that soon after he began his mission in 1861, smallpox forced many of the northern tribes who had gone to Victoria to flee to their homes, leaving their dead and dying along the way. He wrote, "[we] kept vaccinating young and old for days and with the aid of the new Coal Company we succeeded in removing the Nanaimo Indian proper to a new site within their own Reserve, white washing their dwellings, and instructing them in sanitary precautions to prevent any further spread of the disease. In this way they were saved from destruction; whilst, in places beyond us, further north, where no such superintendence and aid could be rendered, whole villages were decimated and in some cases not one escaped" (Good 1861–1900).

[142] Village Island IR#1, Aktis (Arima and Dewhirst 1990: 392) Kyuquot.

▲ Ucluelet Chief identified as MAA-E-kin, c. 1864.

(PHOTO: C. GENTILE — ROYAL B.C. MUSEUM & ARCHIVES: PN-04812)

▲ Two Men in Alberni Area, possibly Tseshaht, c. 1864.
(PHOTO: C. GENTILE — ROYAL B.C. MUSEUM & ARCHIVES: PN-05072)

▲ Quatsino Girl with Conical Head Shape, drawing probably by E.P. Bedwell.
(MAYNE 1862: 277)

▲ Man at Dididaht Village of Whyack, c. 1864.

(PHOTO: C. GENTILE — ROYAL B.C. MUSEUM & ARCHIVES: AA-00882)

▲ Head of Isoris Arm, Nootka Sound.
(MAYNE, 1862: 234)

▲ Eucluelet Village, Barkley Sound, c. 1859, drawn possibly by James Robertson on board the H.M.S. *Satellite*.
(BEINECKE LIBRARY, YALE UNIVERSITY: #1332431)

▲ Comox Village. (ROYAL B.C. MUSEUM & ARCHIVES: G-06607)

▲ Koskimo Village, c. 1870.

(PHOTO: R. MAYNARD — ROYAL B.C. MUSEUM & ARCHIVES: F-08199)

Island and finished sounding the Entrance of the anchorage inside the Village of Markals[143] to the northward of Centre Island and on the East Shore of the Tasis Arm about 4 miles within the Entrance (Gowlland 1861–1864).

JULY 8. After PM sights started and steamed out of Strait of Fuca. On mrng of 9th Sounded about the spot where I supposed I was when 19 fathoms were struck on 18th Aug '61 — occasion of grounding at C Flattery. Could find nothing less than 29 and that on the spot where the 19 was supposed to be. Many canoes fishing for Hallibut — but they were in 52 fathoms. Steered to the westward and Sounded along the edge of the bank in over 100 fathoms until 3 am 10th July when we hauled in for Nootka Sound. At 7.30 am entered the Sound, found our boats here having finished their work westwd. Lay to off Friendly Cove while I got sights. Equal altitudes for the [illegible] Dist between Esquimalt and this place. Gave Mr Gowlland fresh orders and left at 3 PM for Quatsino, carrying the 100 fathom line. Mr Gowlland tells me the Natives here have not been so civil to his parties as they were during the ships presence, and says Maquinna is a very second rate fellow, which I incline to think myself. The Kyuquot Natives appear to have been the most civil and hospitable to our boats. Mr G speaks highly of them and brings me their numbers, represented by sticks, in separate bundles, one for women, one for men, and one for children. The population appears to be ________ men, ________ women, ________ children[144] — (see the sticks in my Canteen).

Fine calm weather all night. Hauled in for Quatsino at 10 am, having passed Woody Pt at 7.30 am. Passed on southd of the Danger reefs — got sketches of the leading marks. 2d bare Surf Id just open northd of Bold pt, until Village Id (North Harb Quatsino) is on[e] with pt of Burnt Mt when [then] steer for the burnt Mt which is very remarkable, on[e]

---

[143] Markale Reserve, Mahqit (Arima and Dewhirst 1990: 392) Kyuquot.

[144] Richards provided the missing numbers in a footnote to his journal entry of September 20, 1860. There he notes that Gowlland told him there were four hundred people. Richards' 1860 estimate noted three hundred fighting men.

half of it having been originally destroyed by fire and perfectly bare of pines. Now grown up with bushes, willow and alder, separated from the pines by a remarkably well defined line. Steering for Burnt M^t^ leads ¾ of a mile East^d^ of Danger Rks and a vessel may round the Surf I^ds^ at a convenient distance and then steer up the Sound.

We stopped for a few minutes off Koskimo Village,[145] where we saw the English schooner Explorer at anchor. She is now chartered for a voyage round the I^d^ by a gentleman named Travers whom we met at Nanaimo a month since. Passed on and anchored at 3 PM 20 miles above the Entrance in a Snug Cove examined by us on a former occasion and now named Hecate Cove.[146] Met Mr Travers in a canoe just before we anchored. He had been to examine the upper part of the Inlet. He did not stay but passed down to his schooner.

JULY 12, SATURDAY. A beautiful day. Got sights and established a Saw pit. A canoe arrived from the Schooner Explorer with a polite note from Mr Travers offering to take letters to Victoria, and to wait our time if we desired it. Sent our letters in half an hour.

QUATSINO

JULY 13, SUNDAY. Warm fine day. The Indians from below came up this afternoon, among them my Cone headed girl, Yak-y-Koss, a friend of more than 2 years standing. She came with her father and mother and expected the usual presents. Gave her a dress and 5 reels of cotton & a packet of needles to make it up. Poor child, her parents evidently look upon her head as a source of great wealth to them, and fancy that there must be some great attraction in its remarkable shape to me. They almost demand potlatch for her. As I was interested in the poor child and civil to her at first, I think it right to carry it out, so I always make her presents.

The only 3 boats left onboard are prepared to start tomorrow for a fortnight — I leave myself for a week. The father of my cone headed girl brought onboard today a large piece of Virgin Copper — and intimated

[145] Maate or Maylattee.

[146] Quattishe.

that it had been found on this Sound. Our ignorance of this language prevented our understanding exactly where it was found. I have my doubts about it being found here, and think it extremely likely that it came from Queen Charlotte I^{d}. He was very careful of it and seemed to think it very precious — which would have scarcely been the case was it found in quantities in this Inlet. He had it wrapped up in a piece of rag and made more fuss about than if it had been gold.[147] Gave my cone headed young woman a cotton dress, 5 reels of cotton, a bar of Soap and a Comb. I like to enumerate my gifts for the information of others, who may come after me. They value soap and I think come onb^{d} dirty to get it.

JULY 16, WEDNESDAY. On Monday the 3 boats left the Ship. Mr Pender and Hankin with 1 Whaler and Cutter to Sound, then Survey the NW arm; myself to examine the one leading to the Rupert Portage. Day not particularly fine. Proceeding nearly 2 miles Eastw^{d} from this cove, we entered the Narrows — an arm running nearly north for 2 miles more and not above 2 cables in width. At its northern end commence the great 2 inlets. The waters from them cause a very rapid tide in the 2 mile Narrows from 6 to 8 miles an hour, with ripplings at the Northern Entrance which would in strong winds be inconvenient to boats. I found that the Eastern arm did not run above 7 miles from the narrows and terminated as all those sounds do in a round head with Swampy ground & several small rivers emptying themselves. The depth of this one is not so great as we usually find — 40 to 50 fathoms ¼ of a mile off its shores, 60 in the middle. There is also anchorage at the head in 12 to 15 a cable from the shore; a creek runs to the SE. On arriving at the north end of the Narrows — where there is good anchorage for small vessels. A vessel of this ships size could lay there comfortably.

The geological formation here is very curious — in some places limestone, close to it again Trap, and in some places the Conglomerate. The lime stone is of a very fine kind — seams of white marble running thro

---

[147] On the Northwest Coast, the "copper" was a costly prestige item, affordable only to those of the highest rank. It was a large, shield-shaped piece, often painted or incised. Coppers were highly valued and much traded among high-ranking people, and, like blankets, could be destroyed as a display of one's wealth. They played an important role in the potlatch system.

it, and it can be had in any quantities close to the water with capital anchorage. I completed my task and returned this evening to the Ship. Found a few natives here from Koskimo Village, 12 miles below. My conical friend returned home when she found I was going away, probably thinking the presents were at an end. I hear every one speak of the ingratitude of these people. They are certainly most indefatigable beggars and the more you give them, the more they ask, and it is equally true that they exhibit no kind of thankfulness whatever, even by sign or look. The one who gives them most, they look to him to get more from; the one who makes a good bargain with them, they appear to have more respect for. This perhaps is a failing not altogether a solely characteristic of savage life, and I am inclined to be less harsh in my judgement. They are savages, the children of nature, and it is human nature in its uneducated state to be selfish and grasping. They see that we have plenty of what they stand in need of, and of course they are very anxious to get it — one is almost led to doubt whether gratitude is not of artificial production.

Mr Hankin who has mixed a good deal with these Indians told me that he had never met with but one who had shewn any gratitude for favours bestowed on him. This was a lad named Friday who we had taken onb^{d} at Barclay Sound, fed, clothed, paid him as a seaman, carried him to San Francisco. M^{r} H had made a kind of pet Servant of him, he went everywhere with him. However at the end of a year he wanted to see his friends and left us — with the understanding that he was to come back in a month. I did not want him. He was little or no use in the ship, but I have always taken these Indians onb^{d} with a view to letting them see our Customs, teaching them English, in fact, for their own benefit. But by some mischance Friday did not join us again at the time agreed on — he found difficulty in getting to us, he said, until a week since we accidentally met him 20 miles off the Coast fishing. He had then been away from us 7 months. He was all but naked. I stopped the Ship — he came up and was so anxious to remain that I made no objection and he was shipped on the old terms. M^{r} H cited this again to me as a strong case of attaching. I said nothing. He soon appeared in M^{r} H's best Shirts and Trousers — nothing was good enough for him, and I believe if he had met his sweetheart after a years absence, he could not have been more delighted. Three days afterwards his eyes were opened. Friday told him early that he had made a great deal of money thro being with us formerly; that he could now pilot vessels into Barclay Sound and that the little English he had learned was a fortune to him. He had come now to perfect himself in the language — nothing else — and as soon as that was accomplished he was off like a Shot. M[r] H was much cast down by what he qualified as the slight ingratitude of Friday, but for myself, I cannot see any thing unnatural in the lads behaviour. All his friends are at Ohiat (Barclay Sound) and why should he stay with us save for his own advantage. There is a degree of honesty and frankness in acknowledging it — we must not look for romance among savages, and I must say these fellows are about the most matter of fact canibols I have ever met. What I most dislike about them is their different behaviour when the ship is present, and when our people are left among them in boats. They have not then the dread of the ships force — and the boats have little to give them. Consequently their demeanour is greatly changed and our boats at Nootka experienced something less than civility during our absence the other day. I spoke to the Natives about this. They were apparently a little ashamed, but after all, we sometimes find similar conduct among more enlightened savages. There is little apparent affection among them. I have seen it exhibited towards their Children, tho there is no doubt that among many of the tribes the practice of destroying[148] the children obtains and did obtain to a great degree a few years since. This too I suppose is a habit which we should get accustomed to if it becomes the fashion.

JULY 25, FRIDAY. Today, having finished all the work above, we moved down to the anchorage in North Harbour at the Entrance of the Sound. After my return to the Ship last week, I finished some work in the lower part of the Sound and on Monday 21 accompanied by D[r] Wood and M[r] Wright, Engineer, went up the No Arm above the narrows to look at the Coal. We found it in a convenient harb[r] on the North side of the

[148] Richards probably means "distorting" rather than "destroying." He is referring to the custom of head shaping which so fascinated him earlier. Previously he refers to the practice of "distorting" the children.

West arm about 2½ miles above the Narrows.[149] The Vein appeared extensive. Cropped out at the waters edge and dipped 15° below the Horizon in a NNW direction about (true). The seam was exposed at the waters edge and was more than 100 yards wide. As it extended northerly, we dug down and found 3 or 4 feet of compact blue clay lying over it — then sandstone & rock then the Coal. Close to this seam we found another tilted up end ways, showing the depth of the seam to be 18 inches at the waters edge. We collected several hundred weight with ease and it burnt very well in a fire we lit. Apparently it was as good as any we had seen at Nanaimo — but on testing it onboard as to its capacity to weld iron, it failed. Still it must be remembered it is only surface coal exposed to the constant action of the water. I am of opinion that when worked it will be quite equal to the best Nanaimo coal. I tried some specimens of the lime stone marble and we returned to the ship on 22^d^. The boats from the west arm rejoined same day. They found that arm extending for 20 miles in a westerly direction. During the remainder of the week we were employed in completing and sounding the lower part of this very extensive inlet — the distance from the sea to the head of the west arm more than 40 miles — to the head of NE and SE arms, 30 miles. The weather has not been very fine during our stay here. Indeed, I fancy that fine weather is the exception of this western coast. The Natives from the Villages below were our Constant visitors. They brought little to barter, and seemed only intent on what they could get as presents. There are only 2 Villages[150] in the Sound, one at the Entrance, North Harb^r^, and the other 7 miles within, opposite to Koprino Harb^r^ & Plumper anchorage. Our anchorage on the north side of the Inlet 18 miles from the Entrance is a very convenient one.

---

[149] Coal Harbour.

[150] The village at North Harbour in Forward Inlet is likely Owiyekumi or Tsootsiola, a Quatsino winter village by the 1860s, later allotted as IR#11 O-ya-kum-la (Galois 1994: 378; Richards1864: 244). Richards' survey chart shows one large rectangular house on the shore in front of Village Islet (North Harbour Quatsino Sound, 1860. UKHO D5348. Shelf Fz). The second village opposite Koprino Harbour is Maate at Koskimo Bay. Richards also refers to a third village, the home of Yak-y-koss, located seventeen miles from North Harbour, probably Quattishe, "which they frequent at this time" (Galois 1994: 358–359; 372).

We tried some of the surface coal obtained in the NW arm and found it would not weld Iron.

SCOTT ISLANDS

JULY 28, MONDAY. At 4 am we started and Sounded up to the Scott I^{ds}, the day not promising very fine and the wind from SE. It improved however and we passed along the South side of the Islands, fixing the positions of the Haycocks and western island very satisfactorily after a 4th attempt. Carried a line of Soundings along the North side and passed down thro the Scott Passage between Cox Island and North Cape. I think there are passages between Lanz I^{d} and the Triangle — at any rate, there appeared to be between Lanz I^{d} and the Eastern Haycock [Beresford Island], also between the two Haycocks and there may be one between West Haycock [Sartine Island] & Triangle I^{d}; but there are several rocks and reefs both west of W. Haycock & East of Triangle, and the water seemed Shoal in this latter passage. A very heavy tide rip as if caused by shoal water extended right across it — a very narrow belt of breaking water which appeared as if the ridge of rocks extended the whole length of the pass. There were tide rips also in the other passages, and I did not feel justified in attempting to pass thro in the ship where so much was at stake. They could not have been examined in boats any time. I have been there but I hope I may be able to accomplish it yet. The Scott Channel is a very good one, 4 miles wide, and I certainly should not recommend the others to be used unless in case of necessity. At 7 PM we shaped a course to pass 8 miles outside Woody pt.

WOODY POINT–ESPERANZA INLET

JULY 29. At 3 am hauled in round Woody P^{t} and at 7 stopped off the Entrance of Nasparti & sent a boat to get angles on Haystack Rock which having completed, we stood out and carried a line of Soundings outside the Dangers off the coast between Nasparti & Esperanza. We entered Esperanza by the Eastern Channel and moored in Queens Cove at 4.30 PM. The day has been wet and unpleasant. Some canoes soon came off from Nuchatlet and after some difficulty I got some Indians to go with Friday, our Barclay Sound Native, to Nootka Sound with a letter acquainting Mr Gowlland with my arrival and directing the parties to return here. They resorted to every device to prevent going — asserted that our boats had left Nootka and gone to Barclay Sound — and so determined were they on this point that even Friday declared it was no use going. However, I insisted and by promise of a shirt & pipe & comb, some bead & tobacco each, 3 fellows started.

JULY 30. This morning Mr Pender with 1^t^ Whaler went to survey Nuchat-let inlet, on the east side of the Sound — on Nootka I^d^ Mr Hankin with Cutter to complete the Soundings of the 2 Channels in, and I went in my boat to survey Eliza Sound. The day not very fine.

> JULY 30, WEDNESDAY. During our stay the Natives paid a formal visit in their large war Canoes; holding 60 men some of them; dress up & Painted Most Extravagantly, and after paddling furiously round the Ship 3 or 4 times chanting their war songs, keeping beautiful time to the music with their paddles they came on board and in form presented Captain R with a very handsome Sea Otter Skin about 7 feet long; and then squatted down on deck patiently awaiting a return present; which they received before making the Slightest attempt to leave; this is their Custom (Gowlland 1861–1864).

JULY 31, THURSDAY. Barometer above 30.30, but the hills covered with vapour, which soon fell in rain. At 8 I left to continue my work, thinking the mist would lift off the hills and that with so high a glass in the best month of Summer it would assuredly clear, but I was mistaken and had to return at 10 am. It poured with rain all day. At 3 PM Mr Gowlland returned, and shortly after the 3 Whaler Mr Blunden, reporting the Pinnace on her way. Friday reached Friendly Cove last eve and Mr Gowlland arrived there an hour after, having completed his work. He started this morning early and reached us in 7 hours, a distance of 30 miles. People all well. They have had a good deal of bad weather. M^r^ G brings me the numbers of a small tribe of about 60 people who live at the head of the SE arm of Nootka Much-a-lat. He describes them a finer made people than any he has met, and less acquainted with Europeans.

They prepared their Arms, Guns and Knives on his approach and met him in [a] hostile way, but he jumped onshore & saluted them with Wakash; in a short time confidence was established and they were very civil. They would not take any of our biscuit until every man in the boats had tasted a considerable quantity before them; evidently they have little if any intercourse with white people. They had a good many skins, which they trade for Blankets with the Nootka Tribe. The Nootka under Maquinna keep them shut up in their arm that they may get their skins for a low price. The tribe being small, they are entirely at the mercy of the larger one. Maquinna was very anxious to go up and kill them, but was opposed by the most influential man of his tribe, the Spouter I have mentioned before. But whenever it suits the convenience or interests of the Nootkas, no doubt they will carry out the design.[151] Maquinna was much annoyed at any opposition being made by his people, and the opposition was made not from any philantrophic motive but because the Spouter makes a considerable harvest of their skins. He bought 7 Martens for 1 blanket while Mr Gowlland was there. These people look upon it quite as fair play to murder a small tribe when it suits their interests. If I can get hold of Maquinna again, I shall instill into him that in the event of his committing the crime he contemplates, his neck will probably come into close contact with the hangmans noose. Certainly it will if I can in any way bring him to justice. This coast ought to be visited occasionally by a vessel of war, and it is a great mistake not to have some of the Missionaries working among the tribes who, from their proximity to the Coasts, would easily be brought under their influence. They only require telling what is right or, at any rate, being made to understand that they

[151] Muchalaht Arm, the territory of the Muchalaht, was highly prized for its numerous salmon streams and plentiful game (Drucker 1951: 233). Living on inside waters, the Muchalaht were cut off from the European trade in guns and ammunition. With the collapse of the maritime fur trade in 1795, the "large outside coast groups were destabilized [economically] and they began to seek rights to salmon streams located along protected inlets" (Marshall 1993: 244). This resulted in warfare and an extended period of conflict between the Mowachaht and Muchalaht people. In the 1870s, the Muchalaht invited the Mowachaht and Ahousaht to a potlatch and peace was made. Throughout the late nineteenth century and early twentieth century, intermarriage brought the Mowachaht and Muchalaht people together and most of the Muchalaht moved to Friendly Cove (Drucker 1951: 231, 234). Eventually, the Muchalaht and Mowachaht officially amalgamated and today are known as the Mowachaht-Muchalaht.

will be punished if they do wrong — and nothing would be so easy as to carry out any threat made for all their villages are within stones throw of where a vessel could get, and they cannot retreat and subsist inland. At this time of the year they are all at the entrances or near the entrances of the Great inlets — as soon as the Salmon go to the Freshwater rivers, there they follow them to their Villages constructed at the heads. Thus Tasis at the head of Nootka will be inhabited by the Middle of August & Friendly Cove deserted. The Salmon fishing time is a very busy one among the Indians. The rule is that the largest river in one of these inlets is assigned to the greatest chief; and the second to the next, & so on. And each Chief appoints a particular spot where each of his dependants is to fish. No natives from other tribes are allowed to intrude. Indeed, this is carried to such an extent that they dont allow a tribe to fish on what they consider their rights outside the inlets at sea for Hallibut and the Red-fish, but demand payment for it. Thus if the Rupert tribe came to fish south of Cape Scott, they first come and pay Blankets to the Quatsinos. Mr Gowlland fancies he has discovered gold in Much-a-lat Arm. He brings some of the sand onboard for examination. The Friendly Cove Indians seem to have an idea of it, and probably this has been instrumental in producing Macquinnas determination to exterminate the small tribe. He, Macquinna, told Mr Gowlland that he was coming to me about the Gold and wanted 200 blankets for permission for white men to dig for it.

JULY 16–19. Muchalat or Giuaquina Arm is a long Inlet running into Vancouver Island. . . . Carried the triangulation some 15 miles up the Inlet, and crossed over on the North Shore to camp; found here a Village; the Natives at first opposed our landing, being all armed with their Musquets and knives; but knowing a little of the language I assured them our intentions were not hostile, and after a time they came more & more Friendly exchanging presents of skins etc., for knives & beads or biscuit. Camped close to their Village. The Name of this tribe is Muchalat from which the Inlet should derive its name. They are fine athletic fellows and taller than most of the tribes on this coast; they number about 80 men women & children; are almost constantly engaged in war with the Nootkas who say their chief Maquinna of a former generation was Murdered by them; and in consequence whenever a fair opportunity offers they Murder as Many Muchalats as possible & the entrance to their inlet being directly in Sight of the Nootka Camp at Friendly Cove if they attempt to leave it are immediately pounced upon; this will partly account for the present savage state in comparison with other tribes; they have had little or no intercourse with Whites and all the trade they carry on is with their Enemies and bug bears the Nootkas; who get enormous profits from them for musquets etc.; we remain camped with them all Sunday and Monday and became very friendly; in return for 3 or 4 deer and several fine skins they presented us with, we returned calico, razors, knives etc. — and they appeared well Contented. The Village is built on the only clear good Bit of land I have seen in the whole sound I should think it is Nearly 1 miles Square, soil good; and very little trouble would clear it, it is situated in a Valley between two high ridges of Mountains forming a very deep Valley through which a large river flows into the Inlet; the Indians say Gold exists some distance up the Valley in the bed of the Stream;[152] probably there is; every appearance of its presence being shown by high boulders and a very fine dark sand (Gowlland 1861–1864).

❧ 1862 ESPERANZA INLET–HESQUIOT

AUGUST 8, FRIDAY. All boats were employed at Esperanza until last evening. Muchalat at the southern entrance is a good harb^{r} in moderate depth of water and clear channel into it. Macquinna & his wife, the Seal, came from Nootka to see us, also the Kyuquot tribe of which Ka-ni-nik is Chief, with his wife Wi-ca-na-nish. The Ehattesat Chief and his wife

---

[152] Gold River.

also arrived at the same time and there was a great Gurherry. Ehatteset is the father of Macquinna & his wife is the girl described in Belchers voyage as Old Maquinnas daughter. She is infirm now & suffering from Rheumatism and other maladies. I presented her and her husband with a mantle of blue cloth each, but she evidently valued a packet of Brick dust which the Doctor prescribed for her complaints at a much higher rate. They gave me a sea otter skin, not a very good one, and made a terrible parade about it. On the 8th Augt we left Queens Cove and steamed up Esperanza Inlet, entering Nootka by the narrow canal which connects the 2 Sounds. Found the water very deep, and the navigation very [illegible] passing between high mountains on either side. Picked up Maquinna and his wife on the way, and landed them in a canoe at Friendly Cove. We then carried a line of Soundings round Pt Estevan and entered Hesquiot harbour [Hesquiat Harbour] or bay at 3 PM. This is a spacious bay, but compared with other anchorages the water is shallow — 10 fathoms at the Entrance, soon shoaling to 7 and 5 with a great deal of kelp growing on many parts. There is a considerable population here, perhaps 200 — and a large Village. They had a good many skins and dogfish oil. I got a Small Sea Otter for 12 Blankets. There was a magnificent one for 40 to be had. We commenced the survey of the bay at once and continued it the next day, with all boats, until 11 am when we left and shaped a course for Fuca Strait, passing Barclay Sound at dusk. The weather very fine. Saw C Flattery light at 10.30 PM and at 8 am of Sunday 10th Augt were off Race Lt Ho. The Ebb running against — 6 kts, it was not until near 11 am that we anchd at Esquimalt, where we found the Bachante. I walked to Victoria and down again. The day proved rainy and unpleasant.

AUGUST 8, FRIDAY. Esquiat, also a Native Name of the tribe . . . number about 300 (Gowlland 1861–1864).

1862 ESQUIMALT

AUGUST 11, MONDAY. We remained in the harbr a week to give the men leave, but on Wednesday at midnight had to go suddenly to sea to endeavour to intercept deserters from Flagship.[153] Ran over and anchored in Ediz Hook or False Dungeness [Port Angeles] or Cherbourg as it is more lately named. Remained there 3 hours, then returned across the Strait. This is a very spacious harb^r, easy of access; anchorage in 7 to 10 fms on South Shore. Open to East but winds dont blow home. Returned to Esquimalt on Thursday morning, found Packet had arrived during the night & on the point of Sailing again. Missed sending some letters in consequence.

> AUGUST 11. Received an order suddenly from the Admiral to get up steam and proceed to lay between the Race Rocks and Port Angeles to endeavor to intercept a boat full of deserters from HMS "Bacchante" who intended coming over this night. . . . The Number of Deserters from this ship laterly has been something alarming, she has already sustained a loss of 100 men; the Cause is not evident; whether undue severity in the discipline Maintained on board; or from inducements held out to the men whilst they are on leave; as the cause of the running away is not as yet known; I should be inclined to think a little of Each. We never lose a man; they have as much leave as they require or ask for, compatible with the rules of the Service (Gowlland 1861–1864).

AUGUST 16, SATURDAY. Rode to Elk Lake.

AUGUST 17, SUNDAY. Spent at Victoria.

AUGUST 18, MONDAY. In morning anch^d off Victoria H^r at 7 am. At 10 embarked the Governor and his Civil staff & towing the Explorer schooner with about 80 settlers proceeded for Cowichin, where we were anch^d at 3.30 PM. Gov & his party landed with the settlers. The object being to establish them on some of the Cowichin land as farmers.

---

153 H.M.S. *Bacchante*.

Anchorage off a village on the south side of the bay,[154] very close in, in 14 fms. Bank very steep. On Surveying next morning we grounded aloft and had to Shift further out.

§ 1862 COWITCHIN

AUGUST 19, TUESDAY. The Governor encamped on what he calls Mt Bruce, an elevation of something like 100 feet, on the north side of the bay. On it stands a Roman Catholic church, the Priest of which seems to have caught more souls — or rather bodies — than his Episcopal brethren. It is a question of who will bid highest for them; if the Catholic Roman is going ahead, it is necessary for the Protestant to launch into more rice or molasses — or to beautify the church or school house a little more. Both do a certain amount of good in checking drunkenness, but as to instilling any principles of religion, I fear, for long to come, this is not to be looked for. They will go to Church and sing and howl as much as may be desired, but they will kill or defraud their neighbours if necessary as soon after as convenient. Schools for the young is the first and only certain means of ensuring success, and then it is a question of years. The Surveyors were employed in marking off allotments of land all today, but the would be settlers shew the greatest apathy and wont even accompany the gentlemen to see the district. A few of them start away with this view in the morning, but as soon as they find they have to walk a couple of miles, they drop off one by one and the Surveyor finds himself left alone.

On Wednesday 20th I was going onshore to visit the Governor and look at the district in the neighbourhood of the River, when he was seen coming off in a canoe. He acquainted me that he had completed his task, and had some interviews with the Natives who were perfectly disposed to receive the white men and allow them to cultivate and occupy any lands other than their potatoe fields and Villages;[155] that the surveyors were going on with their labours and, when completed, every man would if

[154] There are several village sites that are now Indian Reserves in Cowichan Bay. It is likely that Richards anchored off of Kil-Pah-Las, IR#3 or Theik, IR#2.

[155] Between 1850 and 1853 Governor Douglas made treaties with fourteen tribes on Vancouver Island, including those living in and around Victoria, Nanaimo and Fort

he chose be put in possession of an extent of 100 acres subject to the pre-emption law — that is, to occupy and improve.

In the afternoon I took the Governor in my boat to look at the Copper district just this side of Maple bay, on the W side of Sansum narrows. 4 miles dist we found some people prospecting and they had some very fair specimens of Copper ore out, some in Quartz, others in a Calcose slate, the former the most favourable looking. Mr Wigham, an Englishman who has been years working in the Mexican mines, thinks the mount on the East Side, Mt Bruce and Sullivan on Admiral I^d^ are the spots where the Copper will be found, and he thinks the indications here very good. We returned to the ship at 4 PM and visited the settlers who had landed on the S side of the bay. I walked more than a mile inland, with the Governor. The soil appeared very fair but rather light, the ground partially clear or loosely timbered, and no great labour would be required to clear it. The men all acknowledged this but I saw no disposition except on the parts of 2 or 3 to set to and clear and cultivate. After a talk with the Natives and a few trifling presents of tobacco and pipes, we embarked.

VICTORIA–SAANICH

AUGUST 21, THURSDAY. At 4 am we left Cowichan and with a favourable tide passed down the inner channel, anchoring off Victoria at 9 am. After landing the Governor, I steamed into Esquimalt where we remained till Saturday morning.

AUGUST 23, SATURDAY. At 6 am weighed. The packet Oregon arriving at the same moment, I stopped off Victoria till 9 am for our letters and then proceeded up the inner passage of Haro Canal for Saanich. The weather has been remarkably fine and warm during the whole of this month, the heat indeed rather oppressive. On the night of 21^t^ we had some thunder & lightning & rain, which cleared the air and made it much cooler.

Rupert. Under the terms of the early "Douglas Treaties" village sites and gardens were to be reserved and treaty rights to hunt over unoccupied land and to fish as formerly were guaranteed. Compensation for loss of territory was also paid. By 1861 the treaty-making process was over and the Cowichan were not compensated for their lands.

At 2 PM 23d anchd in Coles Bay, Saanich, a very convenient place. I landed at the Village 2½ miles up the Arm,[156] and walked to Mr Lowes the clergyman of South Saanich. He has a small church and dwelling house just under Mt Newton, and close to a farmer named Thompson. I remained here till Sunday afternoon, when I returned to the Ship.

I am much pleased with what I saw of the Saanich district. The country is beautiful, soil very good — in many parts entirely clear, in most loosely wooded with fine Oaks & Maple. But for the great severity of last year, the settlers already established here would have done very well, but not having provided fodder or shelter for their flocks, and the snow lying 3 feet on the ground for 6 or 8 weeks, most of the Stock died. Those however who had any capital at all to enable them to subsist themselves for another year will soon make up the loss. Mr Lowe had a congregation of 29 on the Sunday of my visit. His little church will hold about 80.

1862 NANAIMO

AUGUST 25, MONDAY. 3.30 am we left Saanich and steamed thro Plumper Pass, an ebb of 5 or 6 knots running against us. Water fallen 3 or 4 feet below high springs. It is new moon today. We passed thro at 6.20 am — probably at 5 am we should have had little tide.

At 11.30 am anchored at Nanaimo. Found the HBCo steamer "Labuchere" here, and the Cadboro brigantine, both coaling, so of course there was a difficulty as there always is in this ill-regulated establishment — where every one does as he likes, and there is no one to look after anything. It was only by dint of constant exertion on our own part and sending men to load the coal ourselves, waylaying lighters etc. that we got our coal, 180 tons, by Wednesday evening.

Wednesday afternoon I pulled up the Nanaimo river but the tide not being sufficiently high did not get far. It is pretty, but was very hot and there is no spot where one can land unless in a swamp or on some isolated piece of dry ground where ants, mosquitos and wasps make one glad to get into the boat again.

---

[156] This is currently Coles Bay Indian Reserve IR#3 of the Pauquachin First Nation.

Thursday morning 28th Augt we left Nanaimo at 8 am and went in search of our decked boat Shark which has been employed in Bute Inlet these last 7 weeks. Day very fine and warm. My friend, Dr Benson of Nanaimo, accompanied me for a cruize of 2 or 3 days. Anchored at 7 PM in Oyster Bay just below Cape Mudge in 10 fms. The SE pt of Bay or Kookooshan Spit ⏀ [between] Cape Lazo [and] Mittlenach I^{d} NE, there are 2 or 3 trees on the spit. Cape Lazo is not seen. The 2 inner trees ⏀ the Bluff SE^{d} is the same mark.

AUGUST 29, FRIDAY. Sent Pinnace & 3^{d} Whaler provisioned for 2 months & under Messrs Browning and Blunden to Survey the Entrance of Queen Charlotte Sound and weighed at 7 o'clock, steaming for the Entrance of Bute Inlet. Shortly after 9 saw the Shark and Gig & Mr Bedwell came onb^{d} at 10 am. Provisioned the Shark for 6 weeks and left her to complete the work she was employed on. Sounded down between Mittlenach I^{d} and the continent. At 3 PM a strong SE wind sprung up, very warm and sultry. At 6 abreast Hornby I^{d}, but did not anchor in Tribune Bay as the wind was right in. Continued on slowly for Nanaimo. Wind died after 8 PM. Sultry with Thunder & Lightning. Lay by off Ballinac at 11 PM.

1862 NANAIMO–PT ROBERTS

AUGUST 30, SATURDAY. At 7.30 am anchored at Nanaimo and got 25 tons of Coals onb^{d}. Remained here until Monday morning, 1st of Sep^{t} when we left for P^{t} Roberts to inspect the Boundary monument which rumour said was fast going to ruin from having been built during the frost. At 12.30 anch^{d} off Roberts Spit and landed. Found the Obelisk in Excellent order. On the sea face is printed, cut in large and painted black — the words, Treaty of Washington 15 June 1846. On the Eastern or land side — "Erected in 1861"; on the South, Archibald Campbell, U.S. Commiss^{r}; and on the North side, Capt J. Prevost, RN. Capt. G.H Richards, RN, Lt Col. Hawkins, RE, HBM [Her Britannic Majesty] Commissioners. On the East side also, Lat. 49° 00. Long ________

Saw 2 white men on the ship bound for Yale, and a few Indians taking salmon. They say the fish have not gone up the River this year and are very scarce, but plenty enough outside. Returned to ship at 2.15 and at

2.30 PM steamed for Esquimalt. Many fires burning in the Gulf which gives the atmosphere the[re] a misty foggy aspect. Arrived at Esquimalt at 10 PM. Bacchante here — remained a week. The packet Pacific arrived on Wednesday morning 3 Sep and sailed at 4 PM. Cap Gosset, RE, went by her for England. Mutine arrived from San Francisco Thursday afternoon.

SEPTEMBER 7, SUNDAY MORNING. At 6.30 left for Nanaimo. Carried usual passengers up — Nicols, Skinners. Passed East Point 1 PM, having had Strong Ebb all day. Anchored at Nanaimo at 7 PM. Coaled on Monday and Tuesday. A heavy NW gale blew on Monday which prevented our doing much. The Otter here from Bute Inlet. She left late on Tuesday Eve.

SEPTEMBER 10, WEDNESDAY. Left at 9 am. D^{r} & M^{rs}. B [Benson] and M. onboard. Anchored in Tribune Bay at 2.30 PM.

SEPTEMBER 11, THURSDAY. 4 am left. Abreast C Mudge 10.30. Ebb running. Passed Seymour Narrows at 11.30 and steamed all day to NW. Anchored at Rupert 1 [am] in middle watch.

> SEPTEMBER 11. Arrived at Fort Rupert at 11 PM. . . . Since our last Visit to this place it is sadly changed; the once imposing looking Village in all its rude uncivilized state is now nowhere to be seen. The small pox which went all through the Indian tribes about 3 months ago did not ease off these poor fellows, scores of them died and lay for days and weeks in the same spot unburied and uncared for; until the others through actual dread of their own lives from the fearful efflusion Pulled down the houses of the dead and scattered the pieces found inside, throwing the bodies in the sea to float away as they choose; of the fine muscular, stalwart, fellows that 4 years ago numbered 400 men; now not 50 can be mustered, & they are mostly the Middle Aged or older men, disease appears to have principally attacked the Young and Strong,[157] and those that it treated lightly came to the same sad end through the alcohol supplied them for Spirits by rascally traders. Disease in all its frightful

forms and Whiskey have made sad havoc with these poor unlearned wretches. Wa, wat ley the old Chief is still alive — old Whale dead long ago (Gowlland 1861–1864).

❦ 1862 FORT RUPERT–SUCHARTIE

SEPTEMBER 12, FRIDAY. Got Equals. Landed at Fort. Hankin fell down, knocked my sex^{t} and chronometer over with him and nearly broke his leg — which would not have been half as bad as damaging the instruments.

Our 2 boats had arrived here on 3 Sep^{t}. I left on 5th to survey Q Charlotte Sound. Devastation had arrived on 16th Aug^{t} & left on 18th for Sitka. The Cutter Hambly here, and Schooner Langley (Nicolas Master) on her way. Supposed to be full of whisky for the Indians. Schooner Antelope and Cutter Eagle also arrived from F Simpson, the latter a notorious whisky seller. A passenger in her (Stevenson) came onb^{d} and told me an Englishman was prisoner among the Sabassa Indians[158] 60 miles this side Simpson.

9.30 am today, Saturday 13th, after forenoon sights, left Rupert by the western channel and steamed for Suchartie against a strong westerly wind. Anch^{d} at 1.30 PM and got PM sights. HW at 3 PM — same as at Rupert.

SEPTEMBER 14, SUNDAY. Remained at Suchartie.

❦ 1862 QUEEN CHARLOTTE I^{D}

SEPTEMBER 15, MONDAY. Foggy morning. Started at 5 am, steamed over bar and steered for 17 fathom bank marked by Vancouver NW^{d} of

---

[157] One of the most devastating effects of European contact on the Northwest Coast was smallpox. Acting on a population with virtually no immunity, it had an infection rate of almost 100% (Boyd 1999: 294) and a mortality rate greater than 30% (Boyd 1999: 21). In sparsely populated areas like the Northwest Coast, smallpox was cyclical, appearing every fifteen to twenty-five years as new and non-immune generations appeared (Boyd 1999: 294).

[158] Gitxaala, previously known as *Kitkatla*. However, the incident occurred with the Ginaxangiik, a lower Skeena Tsimshian group. See Richards' entry, September 18, 1862.

Triangle I^{ds}. When within 2 or 3 miles of it, got 41 fathoms, shoaling gradually, then hauled up WNW for S^{o} pt of Q Charlotte I^{d}, the wind freshening from the NW. A heavy short sea [illegible]. At dayl^{t} saw Cape St James bearing W^{d} at 18 miles off. We had been set considerably to leeward during the night. Hauled up at 7 am for the land head to wind. Cape St James slopes gradually from a summit 1000 feet high to the sea. Low at its extreme with 2 apparently detached hummocky i^{ts} lying close off it 180 feet high. Outside these again lie 3 others, one only 100 feet high, bare and whitish. The western side of the Cape is also white faced. Noon in lat 51.59. Steered for Western end of Houston Stewart Channel. 2 remarkable white stripes 6 or 7 miles to the NW^{d} down the sides of the mountains are excellent land marks. The land is covered with pines, but stunted after the Vancouver I^{d} trees.

12.40 passed Antony I^{d} which instead of being 17 miles from Cape James as the chart shews is 12 or more. Inskip[159] says it is not 10. St. Antony I^{d} is covered with trees. It is 200 feet high with white cliffs. Passed on between Bare I^{d} & Gordon Isles — 2 rocks bare lie off the latter extending into the passage which is ¼ of a mile wide. Bare I^{d} is along table top^{d} [island]. On it NWN is a remarkable square rock, like a large Barn or Haystack. The I^{d} is 50 feet high. Anch^{d} at 2.15 PM in 16 fms (hw) in Raspberry cove, just off the stream.[160] Mr Inskips plan is of great service. Got PM sights. HW at 3.30 PM. A few natives came off.[161]

> SEPTEMBER 16. Several Indians Visited us from their Village which appears to be situated on the Island on the Port Hand Entering (Anthony Island). They are of the Northern race; round big heads, with High Cheek bones and broad lower jaws; Cunning, clever looking eyes and great scoundrels gener-

[159] George Hastings Inskip.

[160] In Houston Stewart Channel, off the south end of Moresby Island.

[161] This is possibly the village of Ninstints which is now part of a Canadian National Park Reserve and a UNESCO World Heritage Site. Ninstints was a village of the Kunghit Haida, who died in the hundreds in the 1862 smallpox epidemic.

ally, the traces of smallpox is visible amongst them yet; but the height of the Malady has passed, hundreds of them have died (Gowlland 1861–1864).

SEPTEMBER 17, WEDNESDAY. Fine cold morning. 48 natives came alongside in 4 Canoes. A fair portion of them women & children. Name of tribe here Kitighwalls. 9 am steamed out of East Entrance NE & E for 8 or 9 miles, then to the NW. Passed outside 2 small bare Islets, then 2 wooded ones. One or two small rocks 6 or 7 feet above hw lie a mile or less off shore. Kept a mile outside there — or as NW 2 small wooded Copper I[ts] to the NW just on port bow. Soundings 56 fms. The coast seems much indented and is tamer looking & appears to offer more shelter and Bays than the W Coast of Vancouver. 2 indians accompanied us from Raspberry cove. The elder a very respectable looking man with a brand new bell topper covered with crape. His wife took the hat from its case & brushed it before she gave him up to us. 2 PM anch[d] among a group of small i[ds] in 9 fms, on one of which the copper people are working a shaft. There are 7 white men here, a M[r] Poole at the head of them — a great number of Indians attracted also by their presence have congregated from various parts of the I[d], and naturally been troublesome among so small a party of whites. M[r] P tells me his sledge hammer & some Kettles have been stolen and that the natives have constantly demanded potlatch from them.

Cleugh is the name of the chief (a kind of Bogus fellow, to use M[r] Ps expression), whose tribe have been most troublesome — but as M[r] Cleugh came onboard with M[r] Poole and seemed to be in no wise disconcerted at the ships visit, I do not imagine that his conscience can be very guilty. Kan-skin-e, fellow in green blanket, also came onboard. M[r] P speaks favourably of him and says he assists the white people. Skil-ka tees, the chief who has been most troublesome, is about 10 miles this side of Cum-shewas harbour.

I assembled the people onb[d] and gave them a lecture and a tame shot and shell practice. The latter appeared to have the most effect. Then I visited the Copper place. I dont think it promises much. The white people seemed to be living in very friendly terms with the Indians, particularly with the female portion of them — with whom domestic relations certainly appeared to exist — and I saw nothing to lead me to believe that there was any danger to be apprehended to Mr Pooles party. I desired him to report how things were going onbd the schooner Rebecca, shortly expected, and told the Chief Cleugh that the articles stolen would have to be recovered, or we should visit him again in a less friendly disposition. He promised he would do his best.

The anchorage ½ a mile from the Copper Id is good enough as a stopping place, but there is a good sheltered harbour 2 miles to the SEd of it; at 5.30 PM we left and shaped a course for Bonilla Id 70 miles off on the continental shore as old Vancouver would have said. Soon after leaving the land, the Soundings decreased from 60 fms to 20 on a bank of gravel, and we carried that depth for nearly 30 miles, when it deepened to 70 and 100.

> SEPTEMBER 17, WEDNESDAY. At 11:30 Weighed and steamed out of Raspberry [Cove]. Shaped a course to the Northward along the Coast of the Island, and at 1.30 PM anchored in 9 fathoms in Skincuttle Harbour abreast the Copper Mining Company's Establishment. Mr Poole the Civil Engineer in Charge of the operations came on board and made his Complaint against the Indians; a great Number of the latter were present ranged round the Quarter deck; he accused them of Stealing 7 or 8 Sledge hammers etc. and Various Cooking utensils; which Accusations they of course most Virtuously repudiated, blaming the thefts on some men of more distant tribe; who had left some time ago; the further we enquired into it, the more complicated the matter appeared to pick to pieces; so the Captain gave them to understand if the King George Men were ever Molested again a Man of War would instantly come and Burn all their Villages, and otherwise most seriously punish them the offences; they appeared much amused at our threats; and paid little or no heed to any warnings given; they are not easily intimidated; the only mode of treating indians I find is

> the most decisive measures thought of on the spot and very severe with instant execution; then if they cannot kill you they will in the future fear and respect you (Gowlland 1861–1864).

SEPTEMBER 18, THURSDAY. At 3.30 am saw Bonilla I^{d}. It is remarkable, about 1000 feet high, rising to a dome or nob shaped peak in the centre and a flat-topped hill lower at the northern end. Passed 3 miles on its western side and steamed in for what we considered Ogden Canal, intending to get into Chatham Sound & to reach Fort Simpson in smooth water. After some little embarrassment we got into the sound, but the chart was but little guide. Steamed all day picking our way among islands and small bare rocks. Passed Mr Duncans new mission Station[162] at 2 PM and at 4 passed Finlayson I^{d}. Stood across to Birnie I^{d} to sound a long Shoal or reef which extends off the NW end of the former, and anchored off Fort Simpson where we found the Devastation.

Heard from Cap Pike that he had succeeded in getting 2 of the Indians implicated in the late murder of 2 whitemen near this place by surrounding the Village and Capturing 8 Hostages. There were still 2 other natives to be captured, and Cap P had threatened to destroy the village and property if they were not forthcoming by Friday even^{g} (tomorrow). The tribe (Kin yan Gecky) had all taken refuge on Dundas I^{d} 12 miles west^{d} with expectation that the threat would be carried out — except the Chief, Alam-lah-hat, who was too old and infirm to move. He had been captured & taken onb^{d} the Devastation but Cap Pike had been induced to release him in consequence of his miserable condition.

> SEPTEMBER 18. This place is an Establishment of the HB Company's for their fur trading purposes and is the principal and largest on the Coast, and their furthest North Establishment, the place is called after Sir George Simpson, late Governor of the Company, who chose the sight during his visit to BC some

[162] Old Metlakatla, William Duncan's mission community, south of Fort Simpson.

years ago. The Building is much like fort Rupert, all the Dwelling Houses, Stores etc. surrounded by Strong Wooden palisades 20 or 30 feet high with Bastions at each angle to afford protection to the inmates if attacked by Natives who are very numerous about this District. The Number of Murders and assaults committed by Indians on white men laterly, since gold has been discovered at Stickeen have reached a most alarming figure; no number of miners now under 20 are safe from the Attacks of Indians; and even so large a body require to be well armed and Constantly on the Alert. The route from Vancouver I[d] to the Stickeen Gold fields lies first from Victoria by the inner waters via Johnston Strait to Fort Rupert & up to which place they are perfectly safe from any fears of Indians; but after leaving that place and Entering the Intricate channels adjacent to the Coast of BC; of which country little is known and rarely Visited except by the periodical trips of the HB Company's Steamers; then their troubles begin. They have then a long Voyage of 400 miles before them, and territorys of hostile, powerful, daring Indians to pass through with perhaps nothing but a crazy old whaleboat to carry them through and imperfectly armed. The Belakoula Indians frequent the Shores of BC and Islands Adjacent to Benctinick Arm. They are great rascals and in large War Canoes holding 40 men with perhaps 100 muskets prowl about like pirates ready to pounce on the first unsuspicious Schooner or boat they come across. Gold is a wonderful incentive to induce men to go thro so many dangers to get at it. The "Devastation" is here now with 40 poor fellows who have thrown away all this last year looking for the Scarce commodity of the Stickeen river and who are only to glad to get a passage back to Victoria again in her; not having even seen the Colour; she is also enquiring into the facts of a Murder of 3 white men by 3 men of this tribe the Sim-Sea-Ans that happened some 3 or 4 months ago. Commander Pike on the information furnished by one of the Survivors; and hearing that the Murderers were on the Village onshore; at Daylight the Morning Arrived had Moored and Armed his boats & Surrounded the Village; but only succeeded in capturing one Man and a woman, the other two Escaped. He Seized on 6 of the principle Chiefs of the tribe and brought them on board as Hostages; during the day another of the Murderers was given up; but the greatest Scoundrel and prime mover in the transaction managed his escape to Dundas Island; whither the Captain Dispatched Captain Pike of the Devastation to look for him but after a fruitless search and a Skirmish with the Indians she returned on the Evening of the 20th towing about 20 Canoes they had captured and one prisoner who Nearly Shot the boatswain. Monday 22nd Sept. Captain landed and harangued, tho Mob threatening, that if within 48 hours the Missing Culprit was not given up he would burn the Village; the threat had a good effect for parties instantly started in large Canoes to hunt him up; and they all began packing up and Vacating the place. The Village is very large and built more after the European Style than any other indian houses on the Coast; the tribe originally number 2000 men; but the late smallpox and disease incidents on intercourse with white men have brought them down to Half that Number — they are bold, Strong fine fellows, shrewd & clever and unlike the Vancouver's Indians scrupulously clean; an enemy not to be despised. They threatened that if Sickness had not brought them so low; they would have towed the "Devastation" on the beach "and Made a bonfire of her". Captain McNeal is the Chief Factor in Charge of the trade on this part of the Coast and has been a long time some 30 years at it; the Indians all fear and respect him. Small pox has made dreadful havoc amongst these Savages; 600 of them out of 2000 have already died; and even now it is raging in the Camp; numbers of their houses we see shut up and Marked. They are full of the unburied dead in every stage of decomposition and who can help it? What can be done? If the Companys men in the fort offered to bury them the Indians would not allow it; and they are afraid to do so themselves. Mr Duncan the Missionary who formerly lived amongst them has now little or no Authority here since he shifted with his flock to

Matlacatlah; poor wretches they are much to be pitied. (Gowland 1861–1864).

SEPTEMBER 19, FRIDAY. I visited the village today, accompanied by Cap Pike and Mr Duncan, and had a long talk with the old chief and the other tribes. They said the weather was too bad to get the murderers across, even if they were captured — but 8 of the principal men of the tribes volunteered to go across & I desired Cap P to take them across in the morning and give them a reasonable time to capture the 2 natives. He did so, and returned on Saturday eve having secured 16 canoes and a large amount of winter food together with the women & children who were landed again at the Fort Simpson Village, allowing the men to take one canoe to search for the murderers — whom I think it probable they will never get.

FORT SIMPSON

SEPTEMBER 23, TUESDAY. They brought another of the murderers from Dundas I^{d}, and we landed the armed boats to bring him off. The tribe now wanted to stipulate that we should be satisfied — without getting the 3^{d} man, and hesitated about letting us have this one unless we agreed — but having a strong force landed we marched him into the boat and onb^{d} and there told them that we can not be satisfied without the other man, as there was a rumour that some of this mans friends & family were down about Pearl harb^{r}, 5 or 6 miles southward. I got under weigh & taking Cap Pike and his boats with me, also Mr Duncan, I went down. Anch^{d} in Pearl harb^{r} at 6 PM, and next morning at 5 am went into By bay where we surprised 9 more of their canoes and sent them to Fort Simpson. The people, except one or two old women, had concealed themselves in the bush and the latter endeavoured to deceive us by saying they had gone to Dundas I^{d} to look after the man we sought. We did not believe them but as it was useless to try and get them out of their hiding place, I steamed across to Dundas I^{d} in the hope if the people who were searching there had got the murderer, that they would bring him off. Then I went to Zayais [Zayas] I^{d} just west^{d} of it and, seeing no natives or canoes, I returned to Fort Simpson. Time was now getting precious. Devastation's coals and provisions were running short. I had still my work to finish on the west side of Vancouver I^{d} and winter had commenced in earnest. We had not had one fine day since leaving Rupert — constant heavy rain and snow down to within 500 feet of the waters edge. It was therefore necessary to decide on some course. A rumour reached us today that the murderer had fled to Naas River, 70 miles northward — at any rate, it was pretty plain that there would be much delay before we could get him. He had powerful friends and relatives who appeared determined to keep him out of our way. After consulting with Cap Pike and Mr Duncan, I determined to take the old Chief Alam-la-lah onboard the Devastation again and hold him as a hostage. To have destroyed the village & 25 canoes now in our possession and all the winter food would have occasioned great suffering to the whole tribe — and would have been perhaps attended with fatal results at the commencement of a severe season — and it did not seem just to inflict so severe a punishment on so many women and children — moreover, the tribe had delivered up 3 out of the 4 culprits, and they assured that they were using their best endeavours to get the remaining one. On Thursday [25 Sept.] therefore I got the whole of the people together, landed 5 boats armed, drew the men up on the beach, and explained what I meant to do and my reasons for adopting the course I had — viz to take off the old Chief and to release the Canoes, their winter food, and spare their village. As the ships would be a few days at Meta-catlah, I hoped that our having the Chief in our possession would induce the people to renew their efforts to capture the man, and that we might still get him and be able to set at liberty old Alam-la-hah and the 5 other hostages onb^{d} Devastation. This was clearly explained by Mr Duncan, as also that, tho we might now go away, we by no means gave up our point of Capturing the Murderer, but on the contrary should return and were determined to have him dead or alive — and that we held the tribe responsible to us for him. The old chief addressed his tribe, upbraided them with not having made any exertion to save him from this disgrace and offered 10 blankets each to any men who would capture him. He was then taken onb^{d} and in the afternoon all the canoes and food were returned. In the eve^{g} the boats of Devastation boarded the Cutter Sherman loaded with whisky for sale to the natives — and destroyed the whole of it. She had onboard 300 gallons of alcohol — which would have manufactured about 1000 gallons of poison. They buy blankets with this stuff from one tribe, and barter them again with another for skins. The natives get madly intoxicated, all kinds of excesses follow. Their women are seduced away from them and can it be wondered at that when the opportunity occurs they take revenge. This spirit dealing puts a stop to all legitimate trade. No one can compete with the whisky seller. He gets a blanket at Rupert for half a gallon of poison worth a shilling, and of course can afford to give six blankets for one the honest trader can exchange for skins.

§ 1862 METACATLAH

SEPTEMBER 26, FRIDAY. Everything busy now. Settled as far as it could be at present. I left at daylight this morning and at 9 am anchored off Mr Duncans place at Metacatlah — in a very fair bay sheltered from all but westerly winds, and not an unsafe place even with these. I visited Mr Duncans mission in the afternoon — he has about 500 people here from all the tribes. He is building a large school house and villages for the people. He is going to civilize them, teach them to cultivate the soil and manufacture articles for sale and, in short, to regenerate them. He has about 200 children of both sexes whom he is teaching to read and write, and to abandon their heathen practices — and he certainly has succeeded so far most wonderfully. Many of the people speak a little English and some can write very well, and there is a most marked difference both in their appearance and manners. M^r^ Ds influence over their young and old is most remarkable and, if his life is spared, he seems in a fair way of realizing his scheme. About 300 of his natives came onboard on the day of our arrival — and went round the ship all well dressed and remarkably well behaved. I had about 80 girls between 6 & 10 years old in my cabin. They sat on the deck and sang sacred music in English and their own tongue, led by Mr Duncan. On Saturday we landed at his place and the people had a feast of rice and molasses. We sent them some bread and rice and 8 large plum puddings. The latter were divided among 140 children. During Friday and Saturday we made a plan of the anchorage. Sunday was a strong SE gale and continued rain. Mr Duncans Indians sent me a petition written in English and composed by one of themselves baptized Samuel Marsden, praying I would release their old chief Alam-la hah, as they believed he would die if he were taken away — and promising that they would use their utmost endeavours to apprehend the murderer and give him up when a ship came for him. I was much of their opinion as to the old man dying, and was not sorry to have a good reason for leaving him behind as I directed Cap Pike before leaving Metacatlah — to surrender him to M[r] Duncan, but to retain him till the last moment in the hope that it might renew the efforts of the men still on the search.

SEPTEMBER 29, MONDAY. We left, a strong SE wind blowing. Passed down Chatham Sound and into Grenville Canal and anchored at 6 in a small bight on the north shore about 60 miles below Fort Simpson. Grenville Canal is a deep and almost straight channel bounded on either side by high mountains rising abruptly from the sea. Their summits generally are rock but wooded at their base, down their steep sides rushed innumerable Waterfalls, some of them very grand. The width of the canal under a mile, sometimes not above 1½ or 2 cables across. Our anchorage was in 17 fms. There is a small wooded Islet at the entrance of the place, just west[d] of it and within is a bare rock. The anchorage is outside this rock, about a cable and a half in the centre of the bight.

SEPTEMBER 30, TUESDAY. SE wind and rain. Left at 5 am and continued our course down the channel. Found the chart sufficiently correct to guide us and at 3 PM anchored in 16 fms in Carters Bay of Vancouver. His boats visited it 69 years ago and buried a man of that name who had been poisoned by eating muscles. Saw a spot which might have been his grave. In fact, there was no other piece of soil on the bay. Picked some ferns off it. I dont think there is room to build a hut on the shores of the whole of Grenville Canal, so steep are its sides.

OCTOBER 1, WEDNESDAY. At 4.30 am resumed our course and at 7 entered Millbank Sound, not very well represented on the Chart. Kept along the centre or towards P[t] Day side. Passed northward of a wooded islet, a rock breaking a mile south[d] of it. Then saw 2 bare rocky i[ts] which we passed eastward of a short mile and steered ESE for the entrance to the inner waters again. Cape Swain, the south point of Millbank Sound, is low and wooded, with a detached islet close to it. At any rate, it appears to be detached. It is ill named and should be point instead of cape. As usual, the rocks and islets in Millbank Sound are multiplied. We steered in easterly for 15 miles, hoping to enter the north end of Fishers Canal — but as it appeared choked by reefs & rocks, we turned back 6 or 8 miles and steered down another opening, winding our way southward among hundreds of wooded islets & trying to find a passage thro them eastward into Fishers Canal. At last by turning north and almost retracing the distance we had come, we rounded another I^{d} and at last got hold of the Mainland. We lost so much time by this manoeuvre that it was impossible to reach Safety Cove of Vancouver before dark, so we tried to find a place to drop an anchor and succeeded at last in a small bight on the eastern shore of Fitshugh Sound just north of the Entrance to Deans Canal. 2 small bare rocky i^{ts} lie just north of this cove and a larger wooded one. The anchorage is merely a stopping place just inside a kelp patch which we nearly ran on in the dark; there is 9 feet of water at low. We anch^{d} in 10 fms. I only recommend it as a Stopping place for a night — it is probable there are other such. We had 6 fms in the starb^{d} chains after anchoring, so the bottom is very uneven.

GOLETAS CHANNEL–CAPE SCOTT–KLASH KISH

OCTOBER 2, THURSDAY. Left at 5 am, very fine morning. Steamed down Fitshugh Sound, saw the Pearl and Virgin rocks. They are pretty correctly placed — the former covered or almost so at high water but are a cluster at low. I think one of them may be awash — the Virgin. I only saw one rock about 15 feet out of water. Passed a short mile from Egg I^{d} which is high, wooded and about half a mile long, something the shape of an egg cut in halves longitudinally. The rocks dont extend far off it as shewn on the chart, but Hanna rocks break, bearing about 4 miles SW of the I^{d}, not SSW as shewn on the chart. From a mile off Egg I^{d} the Course for one tree passage is SSE¾E, Egg I^{d} and all the land about Fitshugh Sound is 5 miles, too far west on the Chart relating with Vancouver I^{d}. Sounded in from 56 to 70 fms, steering that course down. At 2 entered one tree Channel [Shadwell Passage] and anchored near one tree I^{d}, a small green I^{d} with one tree on it, in 10 fms. Saw [illegible] just outside the kelp, which grows in 6 fms off Hope I^{d} shore. The day remarkably fine with light southerly wind until 5 PM, when it clouded over. Glass fell and it blew hard from ESE by mid^{t}. At 4 am 3^{d} Oct^{r} —

a heavy gale — let go a 2^{d} anchor, but the flood tide setting in from the $north^{d}$, or rather $west^{d}$, kept her swung beam on to the furious squalls — yet there seemed no great strength of tide. Struck lower yards & top-s^{ls} at 8 am. The glass then 29.45 its lowest. Rain then came on and the gale soon broke. Swung to ebb at 10 am — it blew heavily all day however and rain continued. I consider this the best anchorage in the neighbourhood. Suchartie is small, deep and you must anchor on the steep edge of a bank. Bull $Harb^{r}$ is intricate to enter and small when within, though perfectly land locked. There is plenty of room in this place and we held on well. A considerable tide would set thro at springs. The wind moderated in the night and glass began to rise. This gale was felt heavily at Esquimalt.

OCTOBER 4, SATURDAY. Baro 29.60 at 5 am. Weighed and passed out by the northern channel. Carried a line of Soundings round Galiano & Hope I^{ds}, and towards Cape Scott. A light NW wind and fine weather. Scarcely any swell from the easterly gale of yesterday. Looking for a bank said to exist 6 miles east of the Cape and 2 or 3 miles offshore. It does not exist, but shoal water extends more than half a mile off the Coast 4 miles $East^{d}$ of Cape Scott and kelp extends a long way off, ½ mile quite. Vessels should give C Scott a berth of a mile and the NW shore of Vancouver I^{d} the same. Found a very heavy swell rolling on the shore $South^{d}$ of the Cape. Carried a line of Soundings to entrance of Brookes Bay and into Klash-kish. A very heavy confused sea rolling into the bay, surf breaking heavily on the shore and many detached rocks. Bottom very irregular and rocky, by no means an inviting place for a stranger to run for. Passed a mile $south^{d}$ of Ship rock, which always breaks but now a heavy & constant surf on it. Anchored in the little $harb^{r}$ of Klash-kish at 5 PM. The swell rolls in, quite to the entrance and then suddenly brings itself up as if against a wall, the rebound being almost like a tide rip or breakers. Anchorage in 10 fms immediately under a very remarkable sharp-peaked M^{t}. One miserable canoe with a few Indians came off, but had nothing to sell, they have now left their sea coast villages and gone up the sheltered sounds, taking with them the boarding and matting of their houses and leaving the heavy frame work to be again made habitable in Spring. We found the bottom irregular and rocky and had to let go a 2^{d} anchor in the night.

§ KLASH-KISH–NUCHATLETS

OCTOBER 6, MONDAY. Blew hard from the SW all day with heavy surf breaking everywhere. A boat sounding outside, but unable to do much. Ran up the arm in my boat.

OCTOBER 7, TUESDAY. Left Klash kish. I don't admire it. It is an ugly place for a stranger to run for, tho if embayed with a westly gale is a safe refuge with the chart. The bottom of Brooks Bay very irregular, 20 to 30 fms rocky bottom, altogether an uninviting place, but perhaps Klash kish is a better place to run for than Klaskino. I dont recommend either unless for Coaster or in case of necessity. Rounded Woody P^t^ at 8.30 am, a heavy rolling swell. Calm and very misty. Carried Soundings down to Esperanza. Made the NW Nob pt of Nootka I^d^ at 3 PM. Weather very thick. Passed between the Nob pt and a breaking reef, discovered another rock between the pt and reef which narrows the passage to less than a mile. Heavy surf on both reef & P^t^ and heavy swell which as the passage was but barely Sounded made it unpleasant. Picked our way into a good harb^r^ on the North side of Nuchatlet Sound and moored in 6 fms at head of it.

OCTOBER 8, 9, 10, WEDNESDAY, THURSDAY & FRIDAY. 3 very fine clear days. Boats Sounding Entrance of Esperanza. People Cutting a little wood for the Engines.

§ 1862 OCT BARCLAY SOUND–ESQUIMALT

OCTOBER 11. SATURDAY. Left early. Weather came on very thick and prevented my getting into Hesquiot just south^d^ of Estevan pt where I had a few Soundings to complete, so kept on for Barclay Sound. Fog very thick all night, the Soundings our only guide, and by them gained a position as I supposed 2 miles off Cape Beale in 30 fms. Shingle. Got a kedge ready to drop when the fog lifted for a few moments and we saw Cape Beale 2 or 3 miles dist, bearing WNW. Bonilla pt the same distance. Steamed in just before the fog closed in again. Proceeded up Eastern Channel NElyE which clears channel rocks. Little or no fog when a few miles within and at 2.30 we anchored off Cap Stamps saw mills at Alberni. Found 3 ships and a Schooner loading with Spars & Lumber — a great increase of houses and the place evidently thriving.

Remained at Alberni for 3 days. Visited the mills and the river. Cap Stamp is carrying on his operations with great vigour and is sending excellent lumber to Australia, China and the Ports of Chili, as well as Spars.

OCTOBER 14, TUESDAY. Left the Canal and steamed down at 1.15 PM. At 4 entered a thick fog. Dropped an anchor close to the shore of Copper I^{d} in 14 fms, and remained for the night.

OCTOBER 15, WEDNESDAY. At 5.30 am steamed down the Canal, passed Cape Beale and entered Fuca Strait. At 9 am ran into a very thick fog which we kept till 2 PM. 6.15 PM passed Race Rocks and moored in Esquimalt at 7.30 where Mutine and Forward were lying. Heard that the Admiral had left on 18 Septr for Valparaiso.

All our boats returned from detached service on the 16th. We remained at Esquimalt from the 16th till the 26th Oct^{r}, refitting ship, watering, and giving the men leave.

The weather remarkably fine and mild. The driest year I have experienced since we have been here. Fogs have also persisted this year more than they have done since 1857.

> OCTOBER 16. Found during our absence that the Tynemouth had arrived from England bringing the 60 or 70 young ladies for distribution in the Colony: they were landed at Vic! and put into Barracks at James Bay; the apartments formerly occupied by the Soldiers officers (Marines) — Crowds of the other sex met them on landing — and had it not been for the Marine guard; likely enough they would have been walked off with on the Spot: as it is, such importations will not keep on Vancouver Island. . . . They are composed of Governesses; Cooks, House and Nursery Maids (Gowlland 1860–1863).[163]

---

[163] The "Brideship" arrived in Victoria in September 1862.

ESQUIMALT–NANAIMO–NEW WESTMINSTER

OCTOBER 26, SUNDAY. Left for Nanaimo to make a large plan and buoy the harbour. Here we remained until the 1^{t} of November, by which time the work was completed and both channels well buoyed. The weather has been remarkably fine tho cold the whole time.

The coal mines have now passed from the Hudsons bay to a private Company[164] for 40,000£ — they are working a new pit of superior Coal, but the management of the concern is in very inefficient hands and cannot succeed as it ought until some more capable and energetic director is found. All the arrangements for embarking the Coal are bad, and detain Ships unnecessarily. Skows out of repair, not properly loaded, frequently sink along-side a ship. No piers, bad tramways. In short, a poverty stricken concern. An Engineer is required and an outlay of 12,000 £, then a capable manager, and it will pay handsomely.

NEW WESTMINSTER–SAN JUAN

NOVEMBER 1, SATURDAY. Left for New Westminster where we anchored off RE camp at 4.15 PM. Came on rainy and lasted all Sunday. The Town of NW^{r} has increased somewhat, but there are not the same unerring signs of prosperity as at Victoria. A miserable contracted jealous feeling pervades all classes, military included, and a wretched kind of repining that they cannot alter their geographical position, a desire to quarrel with everyone who comes from Victoria, or says a word in its favour. Certainly the contrast is great between the two places. The miners returning with their wealth from the upper country naturally prefer going on to Victoria to enjoy the winter and will not remain more than the few hours the steamer obliges them in the Fraser. The misfortune of New Westminster is that Vancouver I^{d} with its beautiful harbours and pleasant resting places exists. But they do not reflect that but for the Strait of Georgia being sheltered by Vancouver I^{d}, the river would probably be inaccessible except to small steamers — and to them attended with danger & risk. New Westminster must wait its time. It has advantages which will tell, by and bye, but for years to come Vancouver I^{d} must be the place of first consideration. It is perhaps not singular that

[164] Vancouver Coal Mining and Land Company.

the feeling I have described should pervade all classes who are obliged to reside there — military, the Church. All alike, the same petty jealously exists among them all. If a rejoicing in the shape of a public dinner or ball is got up at Victoria, a New Westminster citizen or soldier dare not attend, even tho he may be there by accident. The head of the military at NW [New Westminister] is an incompetent, miserable, narrow-minded creature — otherwise they would be above such petty nonsense.[165]

NOVEMBER 4, TUESDAY. Left Westminster and steamed over the shoals. Buoys all right, shoal does not alter. Anch^{d} at 5.30 PM off Roberts Spit.

> NOVEMBER 3. The town Council waiting on the Captain with an address to him and the Officers much the same as at Nanaimo. Copy of an Address presented by Municipal Council of New Westminster BC
>
> To Captain Geo Hy Richards RN FRAS etc. etc.–Com^{r} H.M.S. "Hecate"
>
> Sir.
>
> The Municipal Council of N.W. being the only representative body in British Columbia; Hail with pleasure the opportunity afforded them, before your return to England; of offering you their sincere thanks for this Zeal you have ever displayed in portraying the Navigation of the Fraser River in its true Colours for the ability with which you have Conducted its Survey — and more especially for Exhibiting to the world an incontrovertable proof of the great capacity of that Noble stream for Commercial purposes; by your bringing HMS "Hecate" of 900 tons register passed our wharves; a fact which refutes all the statements that have been so industriously circulated regarding our port; and should be sufficient to dispel the Misapprehensions created in the minds of foreign shippers; by those who are interested in retarding our progress.

[165] Richards is likely referring to Colonel Richard Moody.

The Value of your Services to this Colony can not be too highly appreciated; the Accurate Surveys of our Coast will be invaluable to unborn generations of Mariners; and Commerce will ever owe you a debt of gratitude.

Although we may never have the pleasure of Seeing you again in this part of the world, we feel confident you will continue to interest yourself in our favour — Your superior knowledge of the country and of its commercial and Maritime capacities will enable the Thousands of our countrymen at home to judge truly of our position both as regards Commerce and Settlements and though your services may be transferred to a wider & More distinguished sphere of duty we trust we may always have the honor of Claiming you as a friend to British Columbia —

We wish you continued prosperity and success in your career and you & your officers carry with you the grateful remembrance of its inhabitants.

On Behalf of the Council Signed (Henry Halbrook) President. Signed by 61 inhabitants (Gowlland 1861–1864).

NOVEMBER 5, WEDNESDAY. Weighed at 7.30 am and before noon anchored in Roche Harb^{r}, San Juan I^{d}. I came here to visit, victual and pay the Marine Detachment. Went to the camp which is prettily situated and in very excellent order under Cap Bazalgette. I never saw anything neater or better arranged than everything in connection with the detachment consisting of about 70 men.

ESQUIMALT

NOVEMBER 6, THURSDAY. 7.30 am weighed, passed thro Spieden Channel and down Middle Channel. Spent the afternoon Sounding in the eastern part of Fuca Strait. At 5 PM entered Esquimalt Harb^{r} and moored. Here we remained till the 10th Nov^{r}, Monday, when we steamed into Victoria Harb^{r} to fire a Salute and assist in celebrating the coming of age of the Prince of Wales.[166] A great dinner was given in the evening at which Governor & all officials attended. I have rarely attended a more rowdy affair.

NOVEMBER 13, THURSDAY. We returned to Esquimalt, where we remained until Saturday 29th. During this time the crews of the ships were employed Clearing the Freight Ship East Lothian of Powder & ordnance stores for Government, and on the Construction of Various works, Shifting over the Hospital Establishment to the old Engineers Barracks. Hecate, Mutine, Devastation & gunboats lying in the harbr.

NOVEMBER 29, SATURDAY. Left for Nanaimo to fill with coal — that expected from England not having arrived.

[RICHARDS' JOURNAL ENDS HERE]

NOVEMBER 30. Arrived in Nanaimo Harbour. Commenced Coaling (Gowlland 1861–1864).

DECEMBER 4. Recd 26 tons of Coal. VII Unmoored shortened in and proceeded out of Nanaimo Harbour for Esquimalt (Gowlland 1861–1864).

DECEMBER 5. Anchored in Cowalitz Bay, Waldron Island. Saturday 6th Passed down through Herron Is straight for Victoria — Reached Esquimalt and found orders awaiting us from the Hydrographer to organize a party of surveyors to remain behind to conduct the Survey of B.C. Mr Pender (Master) in Charge and Mr Blunden (Masters Asst.) and 6 men from the ship. The Complement during the working season to be made up from the flag ship or senior officers ship in port. The Beaver was hired for the arco modicum at a cost of 1000 £ a year. The Home Government granted 3500 £ a year for the work and the Colonial Government — 500 £ in all 4000 £. . . .

---

[166] Queen Victoria and Prince Albert's son, Albert Edward, the Prince of Wales and heir apparent to the British throne.

Returned to Esquimalt on the 14th November — Awaiting the arrival of the Topaze — and our final sailing orders for England.

Paid our farewell visit to Nanaimo, to fill with Coal; and wish all our friends good bye — they presented the Captain & officers with an address — expressive of their regrets at losing us; and thanking him for all his efforts etc — in advancing the welfare of their little town etc — which he answered by assuring them he had only done his duty and should always watch with interest etc the advancement of Nanaimo etc. — and we parted from Smokey old Nanaimo with Many regrets — and more good wishes for our prosperous voyage — and speedy return. . . .

Arrived at Esquimalt on the 6th. In Evening found the Mail not yet arrived. Hankin & Bedwell rejoined from their Shooting Expedition having slaughtered some 5 or 6 dozen ducks and geese.

Our orders not arriving by the steamer and the Topaze still absent prevented our leaving the island as soon as we hoped. Great deal of talk in the town about giving us a public Entertainment before leaving — Heard from the Hydrographer suddenly on the 7th Decr that a party were to be detached from the Ship and continue the survey under the Command of Pender: Bedwell was offered the Chance and would not accept it, much I think to the Captain's displeasure; I at first would not have remained on the island for any consideration; but reason subsequently occurred which materially altered my views on the subject and for her sake I sacrificed my fondest wishes to return home; and would have volunteered to remain only had not some friends willed it otherwise.

The old Beaver was hired from the Hudson Bay Company for the Service, and towed round to Esquimalt where we commenced fitting her for the peculiar duties she will have to perform; Blunden remains with Pender and six men from the

Ship who all better themselves by getting first-Class petty officers rates — Topaze arrived about the 15th Decr from San Francisco & "Mutine" sailed for the Coast of Mexico to pick up Freight. . . . The Citizens of Victoria offered the Captain and officers a ball on leaving the Colony to commemorate the Event and as a slight token of our popularity amongst them; but he for some reason refused and rather offended in Consequence; the Legislative Council of V.I. present him however with an address; similar to those we rec^d^ from Westminster & Nanaimo; recognizing our services, and thanking us for our labours towards the advancements of the Colony during the past 6 years — accompanied with a very nice letter from the Governor to the Captain; the whole of which appeared in the "British Colonist" newspaper, together with a sketch of our proceedings in the prosecution of the Survey connected with the Island which I furnished the Editor from my journal.

I parted with many regrets the old place where we have all spent such a pleasant slice out of our lives. Making a great many friendships which will doubtless last all one's life — I may say without flattering ourselves too much that no ships officers were ever so universally liked and popular; or no ship ever left so many regrets behind. . . . I trust someday to see the old Island again (Gowlland 1860–1863).

DECEMBER 22. Sailed at 7:30 am from Esquimalt Harbour all the ships cheering us as we passed them steaming out; poor Dan Pender and the Waterman accompanied us to Albert Head towing the old 2nd Whaler my boat for two years in which I had gone over safely so many miles of ground in on the West Coast. They left us at Albert Head not a dry eye in the boat amongst them, and attempted to give us 3 feeble cheers choked with sobs — but it was a complete failure — poor fellows it seemed so hard to part with them — they staid and watched us untill we steamed away out of sights around the Race Rocks, and then pulled slowly back; Blunden wrote me word afterwards that there was not a dry eye amongst them on the boats return — We steamed away round Race Rocks — passed Cape Flattery in the afternoon and at dusk the shores of the old Island faded faded gradually from our view with the approaching nightfall until we lost sight of it all together, maybe for years and may be forever!! — Good bye old Vancouver — & to tomorrow our voyage to dear old England.

— Homeward Bound!!! — (Gowlland 1860–1863).

APPENDIX A

Crew Members of H.M.S. *Plumper* and *Hecate*

	***Plumper*, 1860**	***Hecate*, 1861–1862**
Richards, G.H.	Captain	Captain
Mayne, R.C.	First Lieutenant	First Lieutenant
Hand, H.		First Lieutenant
Moriarty, W.	First Lieutenant	
Hankin, P.J.		Second Lieutenant
Bull, A.	Master & Senior Assistant Surveyor	
Pender, D.	Second Master	First Master & Senior Assistant Surveyor
Bedwell, E.P.	Second Master & Ship's Artist	First Master & Ship's Artist
Gowlland, J.T.E.	Second Master	Second Master
Browning, G.A.	Second Master	Second Master
Blunden, E.R.	Master's Assistant	Master's Assistant
Wood, C.B.	Surgeon	Surgeon
Campbell, S.	Assistant Surgeon	Assistant Surgeon
Brown, W.H.J.	Paymaster	Paymaster
Croker	Assistant Paymaster	
Brockton, F.	Chief Engineer	
Wright, C.		Chief Engineer
Sulivan		Midshipman

APPENDIX B

Index of First Nations

In the course of their survey work around Vancouver Island, George Henry Richards and his crew met numerous First Nations. Each of the nations they encounter is named in the table below. On a few occasions Richards notes the presence of a group while recognizing that they lived somewhere else. For example, he writes that the Nah-kok-toks were fishing halibut at Newhitti, but actually came from the mainland coast. There are other instances where Richards only provides regional population estimates without naming specific groups. In these cases, we have provided names based on known geographical and historical information. Only once does Richards refer to a group whose present day affiliation is unclear. This is for the people he calls "Ao-quas." Our best assumption, based on their location as recorded by Richards, is that they were a group from around Esperanza Inlet that joined the Ehattesaht.

There are numerous First Nations around the Island that neither Richards nor Gowlland mention. The focus of the 1860–1863 survey work was north of Barkley Sound on the west coast and north of Comox on the east coast of Vancouver Island. Only a few of the Nations south of these points are noted. However, for the regions surveyed, most Nations are included. Two notable exceptions are the Huupacasath and Tla-o-qui-aht, although Richards does meet the chief of the Tla-o-qui-aht, Wicannish, on a few occasions at Nootka Sound. The table below and the map showing the locations of the First Nations in the journal include only those discussed by Richards and Gowlland. The location of the encounter is noted, followed by the present day name of each Nation.

Northeast & Northwest Coasts of Vancouver Island: Quadra Island to Quatsino Sound

Name Used in Journal 1860–1862	Location Noted in Journal 1860–1862	Present Name
Eukaltah, Ughcultas, Yucultas	Cape Mudge, Discovery Passage	Laich-Kwil-Tach, Lekwiltok
Matalpie, Matilpir, Mort-teelth-pat	Port Harvey, Call Inlet	Ma'amtagila
Kloitzers		Tlowitsis Nation
Malillacolla, Mah-mah-lillicullas	Broughton Archipelago	Mamalilikulla-Qwe'Qwa'Sot'Em
Knights Inlet Tribe		Da'naxda'xw First Nation
Knights Inlet Tribe		Awaetlala First Nation
Nimpkish, Cheslakees Village	Nimpkish River	'Namgis First Nation
Quogellas, Coquells, Cocquells, Fort Rupert Indians	Fort Rupert	Kwakiutl Indian Band
Quee-peer	Fort Rupert	Kwakiutl Indian Band
Quah-quolth	Fort Rupert	Kwakiutl Indian Band
Wah-lish Quah-quolth	Fort Rupert	Kwakiutl Indian Band
Kom-kutus	Fort Rupert	Kwakiutl Indian Band
Loch-quah-lillas	Fort Rupert	Kwakiutl Indian Band
Fife Sound	Fife Sound	Tsawataineuk
Fife Sound	Fife Sound	Qwe'Qwa'Sot'Em
Nah-kok-toks	"from the main" but fishing at Newhitti	Gwa'sala-'Nakwaxda'xw
Wells Passage	Wells Passage	Gwawaenuk
Klattle-se-Koolahs, Newhitti Indians, Nawitti	Hope Island	Tlatlasikwala Nation
Ne-Kim-Kle-sellers	Sea Otter Bay	Tlatlasikwala Nation
Koskimo, Koskeemo, Quoskimo	Koskimo Bay	Quatsino First Nation
Quatsino, Quatsinough	Quatsino Sound	Quatsino First Nation
Claskino, Claskimo, Klaskino	Klaskino Sound	Quatsino First Nation
Klaskish, Clash-Kish, Klas-Kish, Klash-Klish	Klaskish Sound	Quatsino First Nation

West Coast of Vancouver Island: Brooks Peninsula to Barkley Sound

Name Used in Journal 1860–1862	Location Noted in Journal 1860–1862	Present Name
Kayuquot, Kayoquot, Kyuquot, Kyouquot	Kyuquot Sound	Ka:'yu:'k't'h/ Che:k:tles7et'h'
E-hasset, Ehasset, Ehassets, E-hash-ets, E-hash-ats, Ehattesat, Ehatteset	Esperanza Inlet	Ehattesaht Tribe
Nuchatlat, Nu-chat-lets	Nuchatlitz Inlet	Nuchatlaht First Nation
Ao-quas	Esperanza Inlet	Likely amalgamated with the Ehattesaht Tribe
Nootka, Nootkas, Friendly Cove Indians	Nootka Sound	Mowachaht/Muchalaht First Nation
Much-a-lat, Muchalat	Muchalaht Inlet	Mowachaht/Muchalaht First Nation
Hesquiot	Hesquiaht Harbour	Hesquiaht First Nation
Ahousats	Clayoquot Sound	Ahousaht First Nation
Ucluelets, Uclulets	Ucluelet	Ucluelet First Nation
Toquarts	Barkley Sound	Toquaht First Nation
Sesharts, Se-sharts	Barkley Sound	Tseshaht First Nation
Oū chŭk li sit	Uchucklesaht Inlet	Uchucklesaht First Nation
Ohiats, Ohiat	Barkley Sound	Huu-ay-aht First Nation

South and Southeast Coasts of Vancouver Island

Name Used in Journal 1860–1862	Location Noted in Journal 1860–1862	Present Name
Songhees, Songhies	Victoria	Songhees Nation
Cowichin, Cowitchin, Cowichan	Cowichan	Cowichan Tribes
Nanaimo	Nanaimo	Snuneymuxw First Nation
Komox, Komoux	Port Augusta (Comox)	K'ómoks First Nation

Other First Nations Mentioned

Name Used in Journal 1860–1862	Location Noted in Journal 1860–1862	Present Name
Belakoula Indians	Bentinck Arm	Nuxalk Nation
Sabassa		Gitxaala Nation
Hyders, Hyda, Haiders	Haida Gwaii (Queen Charlotte Islands) and en route to/from Victoria	Haida First Nation
Kitighwalls	Haida Gwaii	Possible Kunghit Haida of the Haida First Nation
Kin yan Gecky	Dundas Island	Ginaxangiik
Chimpcian, Chimpseyans, Sim-Sea-Ans	En route to/from Victoria	Tsimshian First Nations

APPENDIX C

Biographical Directory

Alam-lah-hat: Alamlaxha, the hereditary title holder of the Ginaxangiik, a lower Skeena Tsimshian group. He was twice taken as hostage: once by Captain Pike and again on Richards' orders, to persuade the Ginaxangiik people to surrender four Ginaxangiik charged with murdering two white men.

Baynes, Robert Lambert: Rear admiral and commander-in-chief of the Pacific Station. He entered the Royal Navy in 1810, served with distinction in the Mediterranean and was appointed rear admiral on February 7, 1855. Two years later he was made commander-in-chief of the Pacific Station and was sent to Vancouver Island in 1858 to maintain order during the Fraser River gold rush. He returned in 1859 during the San Juan Island dispute where, by refusing to comply with Governor James Douglas' request to land marines on the island to oust the Americans, he possibly averted war. Baynes pressed the Admiralty to transfer the Pacific Station headquarters from Valparaiso to Esquimalt, and in 1862 the transfer was made. He was knighted on April 18, 1860, promoted to vice admiral in 1861 and to admiral in 1865.

Bazalgette, George: Captain in the Royal Marine Light Infantry. In 1859, Bazalgette, six officers and eighty men arrived on the northwest coast on the H.M.S. *Tribune*. Here they were assigned to Colonel Moody and the Royal Engineers where they acted as guards during road building activities and provided militia training. In 1859, Britain and the United States agreed to joint occupation of San Juan Island, and in 1860 Bazalgette was given command of the garrison there at Garrison Bay. Bazalgette remained on San Juan Island until 1867 and retired a major in 1872.

Bedwell, Edward Parker: Second master and ship's artist on the H.M.S. *Plumper*. Promoted to first master on the H.M.S. *Hecate* when Pender was promoted to assistant surveyor. Bedwell retired a staff commander in 1870. Many of the drawings from this period, including some used in this publication, were made by Bedwell. Bedwell Harbour, Bedwell Sound, Bedwell Bay and Bedwell Islets are all named after him.

Belcher, Edward: Captain, surveyor and explorer. He was captain of the H.M.S. *Sulphur* when she sailed to the northwest coast in 1837 and 1839. Richards was a midshipman under Belcher's command. In 1852, he sailed to the Arctic in search of Franklin, with Richards as second-in-command. He was made admiral in 1872. Richards named Belcher Mountain on Saltspring Island in his honour.

Blunden, Edward Raynor: Master's assistant. After August Bull died, Richards recommended that Blunden "receive some small surveying pay [as] he has long taken an active part in the work and has become a good draughtsman" (Richards 1857–1862: March 16, 1861). Blunden continued to serve with the survey mission, first with Richards and then as second master on the *Beaver* under Pender. He deserted in 1865 after many years of service. Blunden Harbour, Blunden Island, Blunden Rock and Blunden Passage are named in his honour.

Brockton, Francis: Richards' chief engineer on the *Plumper*. As chief engineer, he returned to England with the *Plumper* in 1861. Brockton Point in Burrard Inlet is named in his honour.

Brown, William Henry Joseph: Paymaster on the *Plumper* and *Hecate* until 1861. Richards named Brown Range on Saturna Island in his honour.

Browning, George Alexander: Served as Richards' second master on the *Plumper* and *Hecate*. Richards wrote that Browning and Gowlland are his "best draughtsmen and are most competent surveyors." Browning later served as assistant surveying officer under Pender and in 1868, upon his return to England, he worked at the Hydrographic Office. He retired a captain in 1893. Browning Passage, Browning Entrance, Browning Islands, Browning Creek, Browning Rock and Port Browning are named after him.

Bull, August: Master and senior assistant surveyor on the *Plumper*. In February 1860, he married Edward Langford's daughter, Emma, and died suddenly on November 14, 1860. Richards named Bull Passage, at the northeast end of Lasqueti Island, and likely Bull Island, after him.

Campbell, Samuel: Assistant surgeon on the *Plumper* and *Hecate*. He rose to fleet surgeon during his career and retired in 1877. Richards named Campbell Bay on Mayne Island, and likely Campbell River, in his honour.

Cheslakees: According to Captain Vancouver, Cheslakees was the chief of Whulk, a village at the mouth of the Nimpkish River at the time of Vancouver's expedition in 1792.

Cleugh: Chief of the Haida people at Raspberry Cove, close to the Queen Charlotte Mining Company operation.

Cooper, James: Appointed British Columbia harbour master on January 13, 1859. Cooper joined the Hudson's Bay Company in 1844 and was captain of the *Columbia*. In 1849, he left the Hudson's Bay Company to settle permanently on Vancouver Island. He was at Fort Rupert with Richards during the Hoo-saw-I incident.

Cracroft, Sophia: Lady Franklin's niece and travel companion when they visited Vancouver Island in 1861. She also supported her aunt's tireless work to determine the fate of Sir John Franklin. Richards named the Cracroft Islands and the Sophia Islets in her honour.

Dodd, Charles: Captain with the Hudson's Bay Company. Arrived on the coast in 1834 and eventually became chief factor of the Northwest Coast.

Douglas, James: Governor of Vancouver Island and British Columbia. Douglas joined the Northwest Company at the age of sixteen and was employed as a second-class clerk in the Hudson's Bay Company after the amalgamation of the two companies. Douglas arrived in Canada in 1819 and moved west to the Peace River country. In 1827 he opened Fort Connolly at Bear Lake and married Amelia Connolly, daughter of the chief factor, William Connelly, and his Cree wife Miyo Nipiy, also known as Suzanne Pas de Nom. In 1830, following a conflict with the local First Nations, he transferred to Fort Vancouver where he eventually became chief factor. In 1843 he began construction of Fort Victoria and in 1846, he was made head of the Company's business on the Northwest Coast. In 1849 Douglas moved the Hudson's Bay Company headquarters from the Columbia River to

Fort Victoria. In the same year, Vancouver Island was granted to the Company on the condition that a crown colony be established. Douglas was appointed Governor of Vancouver Island in 1851 and set out to develop the region for settlement, including making a series of fourteen land purchases from First Nations on Vancouver Island. This included land around Victoria, where British settlers were arriving, Nanaimo and Fort Rupert, both of which had significant coal deposits. In 1856 he established the first House of Assembly. An economy based on farming, lumber, coal and salmon fishing gradually began to attract settlers, but the discovery of gold in 1858 flooded Vancouver Island and the mainland with gold prospectors, precipitating the establishment of the Colony of British Columbia and the appointment of Douglas as governor of both colonies. By 1860, with the assistance of the Royal Engineers, the interior of British Columbia was open to settlement. His final act as governor was to establish a legislative council. He retired in Victoria in 1864 and died in 1877. He is known by some as the "Father of British Columbia."

Duncan, William: An Anglican missionary sent to Victoria in 1857 by the Church Missionary Society of England. Shortly after his arrival, he travelled to Fort Simpson to learn Sm'álgyax, the Tsimshian language. At the fort, he found approximately 2300 Tsimshian people, and within eight months, he was preaching in Sm'álgyax. In 1862, Duncan and his converted Tsimshian followers moved to a Tsimshian winter village site, Metlakatla. Amazingly, the people who moved to Metlakatla escaped the devastating 1862 smallpox epidemic, while those who remained at Fort Simpson suffered terrible losses. At Metlakatla, Duncan and his followers built a self-supporting "industrial mission." In 1887, following a conflict with the church, Duncan and 823 of his followers moved to Annette Island in Alaska to build the second community of Metlakatla. Old Metlakatla then saw a resurgence of traditional culture as the church's influence diminished. New Metlakatla became a "modern Christian industrial town" and Duncan remained there until his death in 1918.

Edensawe [Edenshaw], Albert Edward: A high ranking Haida man from Haida Gwaii who knew the waters well and was regularly used by the Royal Navy and the Hudson's Bay Company as a pilot. Born around 1810, he was named Gwai-gu-unlthin. One of his adult names, Eda'nsa, was anglicized as Edenshaw, by which he is known. In his early life, in an effort to become a great Haida chief, he moved strategically within both the Haida and colonial worlds. In 1852, he was suspected of conspiring to plunder the ship *Susan Sturgis*, and used the opportunity to gain favour among the Haida and Europeans. He was baptized Albert Edward, after Prince Albert, in 1885. He died in 1894.

Franklin, Jane: Lady Jane Franklin was the second wife of Rear Admiral Sir John Franklin, the renowned Arctic explorer. When Franklin and his crew disappeared in 1847, she spent many years sponsoring search expeditions and pressuring the British government to discover the fate of her husband and his crew. Expeditions were made in 1850, 1851, 1852 and 1857. Richards was part of the 1852 Belcher expedition. In 1861, Lady Franklin, accompanied by her niece, Sophia Cracroft, visited Vancouver Island and the mainland. Richards named Vancouver Island's Franklin Range in honour of Sir John and Lady Jane Franklin.

Franklyn, William Hales: Captain in the Mercantile Marine, and in July 1860 he became Nanaimo's stipendiary magistrate, a post he held until 1867. His home, "Franklyn House," was the centre of hospitality and social activity in early Nanaimo. It was built of imported brick and California redwood, and was a landmark in Nanaimo until 1954 when it was torn down to make way for the new city hall parking lot. Franklyn left Nanaimo in 1867, and shortly after was appointed the chief civil commissioner of Seychelles Islands where he died in 1874. Daniel Pender named Franklyn Range in Loughborough Inlet after Captain Franklyn. Sidney Bay and Harold Mountain, both of that region, are named for his two sons.

Friday: Likely a Huu-ay-aht man who travelled with Richards on several occasions, as guide, translator and pilot. Richards took him on board and paid him as a crewman and, in this capacity, he travelled to San Francisco. During this time he learned English which assisted him in piloting ships in Barkley Sound.

Fulford, John: Captain of the H.M.S. *Ganges*, Rear Admiral Baynes' flagship, stationed on the Northwest Coast from 1857–1860. He retired an admiral in 1877.

Gordon, G. Tomline: Colonial treasurer arrested for "defalcation of public money" in December 1861. Prior to his arrest, the Duke of Newcastle, the secretary of state for the colonies, had informed Governor Douglas that Gordon had "carried on a complete system of fraud and swindling" in Britain and that he had been forced to leave the country (Akrigg 1977: 231). He was charged with embezzlement and imprisoned, but escaped from debtor's prison on May 18, 1862, and fled to the United States.

Gosset, William Driscoll: Captain of the Royal Engineers. In 1858, he, with his pregnant wife and son, travelled with Colonel Moody to British Columbia. He served as treasurer and postmaster for the Colony of British Columbia and treasurer for the Colony of Vancouver Island.

Gowlland, John Thomas Ewing: Richards' second master on the *Plumper* and *Hecate*. He led many detached survey missions around Vancouver Island and can be credited for much of the survey work that was conducted by the crew of the two ships. Richards wrote that Gowlland and Browning are his "best draughtsmen and are most competent surveyors." In 1863, upon his return to England with the *Hecate*, he was promoted to Master. In 1865 he was stationed in Australia where he married Genevieve Lord, with whom he had three children. Here, as chief assistant, he conducted survey work along the coast of New South Wales and was promoted to staff commander in 1874. On August 14, 1874, while surveying Middle Harbour off of Sydney, his survey vessel capsized and Gowlland drowned while attempting to swim to shore. He was buried with full naval honours in the cemetery of St. Thomas Anglican Church in Sydney. He was thirty-six years old. In a letter written to Gowlland's sister "Birdie" his brother Richard states that Gowlland "often talked . . . about the perils he had passed through on the sea and often wound up with the words that he could see his way to leave a service which he had no hesitation in calling especially dangerous. . . . [H]e was so bright and alive and energetic that one felt more and more proud of him as one knew him better. His loss will be a perpetual regret to us all" (Phillips and Joscelyne 2007: 36). Richards named Gowlland Harbour, Gowlland Island, Gowlland Rocks and Gowlland Islet in his honour.

Grant, John Marshall: Captain of the second group of Royal Engineers to sail for British Columbia. Accompanied by his family, Grant arrived in November 1858. He remained in the colony for five years, supervising surveys, construction, and road building, including the wagon road from

Hope to Princeton. He eventually became a commander in the Royal Engineers and retired in 1882.

Griffin, Charles John: A Hudson's Bay Company employee in charge of Bellevue Farm, a Hudson's Bay establishment on San Juan Island. Griffin was a central player in what became known as the "Pig War." This incident was precipitated by the killing of a Bellevue Farm pig by an American, Lyman Cutler, who claimed the pig was in his potato patch. Griffin demanded that Cutler pay for the pig. He refused and appealed for protection from the U.S. military. In July 1859, Captain George Pickett of the US Ninth Infantry arrived with 60 men. Captain Richards was sent to San Juan Island and was pivotal in maintaining peace during this incident. The resulting joint occupation of the island by the British and Americans from 1860–1872 is still honoured at Roche Harbour, on San Juan Island every summer evening when the British, Canadian and US flags are lowered and each national anthem is played.

Halbrook [Holbrook], Henry: Mayor of New Westminster (1862–1863, 1867–1869 and 1878). By 1864, he was a member of the Legislative Council of the Colony of British Columbia and, for a short period in 1871, he was appointed the chief commissioner of Lands and Works. In 1872, he became leader of the opposition in Premier Amor De Cosmos' government but lost the election in 1875. Holbrook remained in British Columbia until 1880 when he returned to England due to his health. The flags in New Westminster were lowered to half-mast upon news of his death in 1902.

Hall, Robert: Captain of the H.M.S. *Termagant*. He arrived at Esquimalt in May 1860. After running aground on July 31, 1860, Hall sailed his ship to Mare Island for repairs and did not return to the Northwest Coast.

Hand, Henry: Richards' lieutenant on the *Hecate*. He replaced Lieutenant Mayne in 1861. Hand eventually became commodore and was in charge of the naval establishment in Jamaica. Hand Island, Hand Bay and Hand Mountain were named after him.

Hankin, Philip James: Richards' junior lieutenant on the *Plumper* when she arrived in Victoria in 1857. He left the *Plumper* in 1858, making an overland journey before sailing for England. He sailed back to the coast on the *Hecate*, on which he served as Richards' lieutenant until its return to

England. In 1864, Hankin left the navy to become the superintendent of police in the Colony of British Columbia. Later that year, the navy reinstated him for his actions against the Ahousat Indians after the destruction of H.M.S. *Kingfisher*. He served as colonial secretary for British Columbia from 1869–1871. Richards named Hankin Island, Hankin Point, and Hankin Range in his honour.

Harvey, Thomas: Captain of the H.M.S. *Havannah* on the Northwest Coast from 1855–1859. Richards named Port Harvey after him and Havannah Channel after his ship.

Hoo-saw-I: A Snuneymuxw woman and the mother-in-law of John Dolhott, a Nanaimo settler, who was married to Hoo-Saw-I's daughter. Dolhott petitioned the government for help when his mother-in-law was captured by the Fort Rupert Kwakiutl. Richards negotiated her freedom and Hoo-saw-I was returned to Nanaimo.

Inskip, George Hastings: Master of the H.M.S. *Virago* on the Northwest Coast from 1852–1855. In 1853, Inskip made several surveys, charting the northern waters of British Columbia.

Jefferson: A Haida chief of Skidegate who was taken prisoner by Captain Robson at Willow Point.

Ka-ni-nik: Chief of the Kyuquot in August 1862.

Kan-skin-e: A Haida man at Raspberry Cove who cooperated with the miners of the Queen Charlotte Mining Company.

Langford, Edward: The manager of the Hudson's Bay Company Farm in Esquimalt, often called Colwood Farm. He was known for his mismanagement of money and entertained regularly at the expense of the Company. August Bull, master of the *Plumper*, married his daughter Emma and died a short time later, in November 1860. By then Langford had run the farm into debt and the family left Victoria for good on January 12, 1861.

Lewis, Herbert George: Captain in the Hudson's Bay Company and in charge at Fort Rupert in July 1861. Lewis was captain of the Company ship *Otter*, and other Company boats around Vancouver Island.

Lyall, David: Surgeon on the *Plumper* until 1859. He was replaced by Dr. Wood in 1860. Lyall and Richards had served together under Belcher on the Franklin Expedition. He retired in 1873. Lyall Point and Lyall Harbour are named in his honour.

Maitland, Thomas: Rear admiral and commander-in-chief of the Pacific Station. He arrived in Esquimalt in April 1861, after the departure of Rear Admiral Baynes. Maitland's arrival in Victoria resulted in strained relations between the Royal Navy and the colony. "Admiral Maitland was a rigid martinet, and his flagship a most unhappy one. Deserters began fleeing the *Bacchante* almost as soon as she arrived. . . . At one point twenty of the flagship's men . . . headed *en masse* across the Strait of Juan de Fuca to find sanctuary on American soil" (Akrigg 1977: 221–222).

Maquinna: Mowachaht chief at what is now Nootka Sound. This name has been passed down from generation to generation, resulting in a succession of leaders with the same name. In the 1860s, the man who held this position was an important figure in the local trade economy. It was Maquinna who was called upon to dispense punishment for infractions committed by the Mowachaht against Richards.

Mayne, Richard Charles: First lieutenant on the *Plumper* and *Hecate* and author of *Four Years in British Columbia and Vancouver Island*. Promoted to commander in 1861, he left to take command of the H.M.S. *Eclipse* in New Zealand. He was wounded in 1863 in the ongoing conflict with the Maori in the New Zealand Wars. His 1864 promotion to captain occurred while he was recovering in England, and, by 1866, he was back at sea and collecting fossils of ancient quadrupeds from the Straits of Magellan, on behalf of Charles Darwin as well as paleontologists and geologists. In 1870, Mayne married Sabine Dent. He retired an admiral in 1879, entered politics and served as a member of parliament until his death in 1892. Richards named Mayne Island, Mayne Bay and Mayne Passage in his honour.

McNeill, William Henry: Captain in the Hudson's Bay Company. Born in Boston, McNeill came to Vancouver Island on the *Llama*, an American brig, which he then sold to the Hudson's Bay Company. Upon its sale, the Company hired McNeill to remain captain of the ship. He eventually commanded the S.S. *Beaver*, and later was in charge of several different northern posts.

Marsden, Samuel: Chief Shooquanahd of the Tsimshian and one of William Duncan's first converts. He was named for Samuel Marsden, a well-known

Anglican missionary in New Zealand and Australia. His son, Edward Marsden, was ordained as a minister and later challenged Duncan's authority, arguing that Duncan should be succeeded by a Metlakatlan.

Meares, John: An English merchant sailing under the Portuguese flag who is reported to have purchased land at Friendly Cove from Maquinna in 1788 where he built a small house and installed a cannon. In 1789, Don Estéban Jose Martínez of Spain, determined to uphold Spanish claims on the Northwest Coast, seized two of Meares' ships and arrested James Colnett of the *Argonaut*. Eventually, Spain accused England of violating Spanish sovereignty while England accused Spain of treachery. When Meares arrived in London in April 1790, the British prime minister, William Pitt, used him to rally parliament and obtain royal permission to mobilize for war. In the end, war was avoided and a diplomatic solution was reached in a series of three treaties known as the Nootka Conventions. These ensured the release of British ships and men, the return of Meares' land at Nootka Sound, and equal trading, fishing, navigating and settling rights in the Pacific and South Seas. Damages were also to be paid by Spain, and eventually there was agreement for the mutual abandonment of Nootka Sound, leaving it open for all ships of all Nations to use but to be settled by none (See Gough 2007: 118-126; Twigg 2004: 156–159; Nokes 1998: 123–169).

Meesun: "Miss Meesun" was a celebrated Vancouver Island traveller mentioned by both Richards and Gowlland. No further information was found.

Mitchell, William: Captain and Hudson's Bay Company master mariner who was in charge of Fort Rupert in 1859. In 1851, as captain of the Company ship *Una*, he discovered a gold vein at Haida Gwaii in a bay later named Mitchell's Harbour by Richards. The Haida people resisted the extraction of their gold and took some by force. Upon its return to Victoria, the *Una* was wrecked off Neah Bay. During Richards' time on the coast, Mitchell was captain of the *Recovery*. Richards also named Mitchell Bay in his honour.

Moffat, Hamilton: Arrived on Vancouver Island in 1849 as a Hudson's Bay Company employee. He was first stationed in Victoria and was in charge of Fort Rupert in 1862. In 1863, he transferred to Fort Simpson and then to the interior. He retired in 1873 and joined the Indian Department in Victoria. Richards named the Moffat Islands in his honour.

Moody, Richard: Colonel of the Royal Engineers based at New Westminster. At the age of sixteen, Moody began his career with the ordnance survey of Great Britain. In 1858, after a twenty-six-year career with the Royal Engineers, he was promoted to brevet colonel and a few months later, Colonel Moody accepted the appointment of Chief Commissioner of Lands and Works and Lieutenant Governor of British Columbia. The War Office also made him Commander of the British Columbia Detachment of Royal Engineers. As the colony was to be self-supporting, the expenses of the Royal Engineers were to be defrayed by land sales. Moody chose New Westminster as the capital but he proved inept at managing finances, and costs soared. He was also accused of nepotism toward government officers in land pre-emptions, and Moody himself came to own several thousand acres of land. Nevertheless, he is credited with helping to establish the Royal Columbian Hospital in 1862, and his library became the foundation of the city's public library. In 1863, Moody departed with his wife and seven children, along with twenty-two officers and their families, leaving behind 130 Royal Engineers and miners who chose to settle in British Columbia. He returned to England and retired a major-general on full pay in 1863. He had hopes of returning to British Columbia but died in England in 1887.

Moriarty, William: First lieutenant on the *Plumper*. He returned to England with the *Plumper* in early 1861. In 1866, he was promoted to commander and retired a captain in 1881. Richards named Mount Moriarty on Vancouver Island for him.

Nicol, Charles: Manager of the Hudson's Bay Company coal mine in Nanaimo and of the Vancouver Coal Mining and Land Company (VCML), the company that bought the coal mine operation at Nanaimo from the Hudson's Bay Company in 1862. In 1867, Nicol became a magisterial member of the Legislative Council of the colony of British Columbia.

Parsons, Robert Mann: Second captain in the first group of Royal Engineers to be sent to British Columbia. He was the only member of the detachment who did not volunteer but, due to his skill in surveying and map making, he was ordered by the War Office to accompany the detachment. He arrived at Fort Langley in 1858 and was promoted to captain in 1862. He returned to England with the other Royal Engineer officers in 1863. He retired a major-general in 1879.

Pearse, William Alfred Rumbulow: Captain of the H.M.S. *Alert* who served on the British Columbia coast from 1858–1861. Richards named the Pearse Islands in his honour.

Pemberton, Joseph Despard: Surveyor and engineer for the Hudson's Bay Company. He arrived at Fort Victoria in June 1851 and began to survey around the fort and town site, identifying lots and setting aside areas for churches, schools and parks. In four years he surveyed much of southeastern Vancouver Island. Pemberton also made his mark in politics and was elected to the first House of Assembly on Vancouver Island. In 1858 Pemberton became the colonial surveyor and between 1860 and 1864 was surveyor general of Vancouver Island. On May 28, 1860, Richards wrote "a rotten corrupt surveying department is the Curse of a new Colony." Pemberton's land acquisition practices were controversial and, upon his retirement in 1864, he owned more than 1,200 acres. By the 1880s, Pemberton was a wealthy man, living in a mansion in Oak Bay. As a gift to the government, he designed, surveyed and built Oak Bay Avenue, which ran through his property. Pemberton died in 1893.

Pender, Daniel: Richards' second master on the *Plumper* and, after Bull's death, promoted to first master and assistant surveyor on the *Hecate*. When Richards left British Columbia, Pender was assigned to the *Beaver*, a Hudson's Bay Company ship hired by the Royal Navy at a cost of £3500 a year to the Navy and £500 a year to the colony. He continued with the survey of the coastline of British Columbia until 1870 when he returned to England to work as assistant hydrographer at the Hydrographic Office. He retired a captain in 1884 and died in 1891. Richards named Pender Harbour and North and South Pender islands after him.

Pickett, George Edward: Captain, United States Army. In 1859, Pickett was dispatched along with sixty soldiers to protect American sovereignty on San Juan Island. The Hudson's Bay Company maintained a farm on the island and British and American ownership of the island remained ambiguous. The shooting of a Hudson's Bay Company pig by an American settler and the resulting charges against him resulted in the "Pig War." Governor James Douglas sent Royal Navy ships against Captain Pickett and his sixty soldiers but violence was averted when Richards arrived with orders that, contrary to Governor Douglas' previous orders, the navy was not to engage the American soldiers. Pickett later became famous during the Civil War for "Pickett's Charge" at the Battle of Gettysburg.

Pike, John William: Commander of the paddle-sloop H.M.S. *Devastation*. At Governor Douglas' request, Pike was charged with monitoring the coastal whiskey trade, and in September 1862 he was sent to the Stikine River and Fort Simpson. Here he worked with Richards in an effort to apprehend the four Ginaxangiik people charged with murdering two Europeans.

Poole, Francis: Employee and geologist of the Queen Charlotte Mining Company at Haida Gwaii.

Prevost, James Charles: Captain and senior naval officer at Esquimalt, 1857–1860, commanding the H.M.S. *Satellite*. In 1853 Prevost charted the east coast of Vancouver Island and Haida Gwaii on the H.M.S. *Virago*. In 1856 he was appointed first British boundary commissioner tasked with resolving the dispute over the British-American maritime boundary. Richards was later appointed second commissioner.

Rupert Jim: A Kwakiutl man from Fort Rupert who was employed by Richards as an interpreter. He often provided Richards with information and population statistics about the First Nations on Vancouver Island.

Simpson, George: Governor of the Hudson's Bay Company. He arrived in North America in 1820 and was made governor of the Northern Department shortly after the amalgamation of the Hudson's Bay and North West companies. In 1826, due to his efforts to expand the Company in the north, he was made governor of both the north and south jurisdictions. Simpson played an important role in developing the company's 19th-century monopoly in North America. He remained governor of the company until his death in 1860.

Skil-ka-tees: A Haida chief from the Cum-shewas Harbour region.

Stamp, Edward: Captain in the British Mercantile Marine. He was employed as the agent of a London firm, Anderson and Company, to ship spars (trees for ship building). He started a sawmill in Alberni in 1860 but resigned in 1863. He was involved in many business enterprises, including "Stamps Mill" on Burrard Inlet.

Stewart [Stuart], Charles Edward: Captain in the Hudson's Bay Company and in charge of the company post at Nanaimo until it closed in 1859. He then established his own trading post in Barkley Sound where Gowlland notes his presence on June 2, 1861. He died of bronchitis in December 1863 and is buried in Nanaimo. Richards named Stuart Island and Stuart Channel in his honour.

Sulivan: Transferred from the H.M.S. *Tartar* in 1861 to become Richards' midshipman on the *Hecate*. Richards had served under his father, Bartholomew James Sulivan, on the H.M.S. *Philomel* from 1842–1845. Richards states "I shall do my best to bring him forward and hope he will turn out as good a man as his father" (Richards 1857–1862: May 2, 1861).

Tugwell, Lewen Street: Minister and member of the Church Missionary Society. In August 1860, he and his wife travelled to Fort Simpson to assist William Duncan. They remained until 1861, but due to Tugwell's poor health they returned to Victoria. Richards named Tugwell Island near Metlakatla in his honour.

Vancouver, George: Captain in the Royal Navy who charted the waters of Vancouver Island and coastal British Columbia on the H.M.S. *Discovery* and *Chatham* in 1792 and 1793. Richards uses his charts and makes reference to Vancouver throughout the journal. He named his son, who was born in Victoria in November 1860, Vancouver Alexander.

Wack-la: An Ahousaht chief.

Washington, John: Rear admiral and hydrographer of the Royal Navy from 1855–1863. Richards succeeded him as hydrographer following his death in 1863. Richards named Mount Washington in his honour.

Wa-wa-te: A head man at Fort Rupert, who Richards refers to as an old chief. A man named "Wawattie" placed his X on the "Quakeolth Tribe" Fort Rupert Douglas Treaty, 1851. It is likely that this is the same person.

Waynton [Weynton], Stephenson: A Hudson's Bay Company employee posted at Fort Rupert during this period. He assisted Richards in the successful return of Hoo-saw-I. Richards named Weynton Passage in his honour.

Whale, Old: A head man at Fort Rupert who worked in cooperation with the Hudson's Bay Company. Gowlland refers to him as the second chief. He may be the same "Wale" who placed his X on the "Queackar Tribe" Fort Rupert Douglas Treaty in 1851. He died in late 1861 or early 1862 of an "internal complaint" (Gowlland April 8, 1862).

Wicannish [Wickaninnish]: Chief of the Tla-o-qui-aht people of Clayoquot Sound. This is a high-ranking position, and all trade in Clayoquot Sound went through this chief. The name Wickaninnish was passed down from generation to generation, resulting in a succession of leaders with the same name. Through a series of wars in the late 18th and early 19th centuries, the Wickaninnish of the time expanded control at least as far as Barkley Sound, controlling the trade monopoly for the region. The presence of Wickaninnish in Nootka Sound in 1860 suggests an alliance between Maquinna and Wickaninnish. There was an alliance between their predecessors in 1789 when Maquinna moved his people to live with Wickaninnish following the murder of the second-ranked Mowachaht chief, Callicum, by the Spanish. Wickaninnish Island was given this name on the chart in 1861, likely by Richards.

Wood, Charles Bedingfield: Ship surgeon on the *Plumper* and *Hecate*. When Dr. David Lyall left the *Plumper*, Wood was assigned at Richards' request. He drowned in 1865 in the Mediterranean Sea, with several other officers when their small boat capsized during a squall. Richards named Wood Mountain, Wood Islands and Bedingfield Range in his honour.

Work, John: Joined the Hudson's Bay Company in 1814, retiring shortly before his death in 1861. His posts included Fort Vancouver, Fort Simpson, Fort Victoria and Fort Colville. He later became a member of the Legislative Assembly on Vancouver Island, a position he held until his death at age seventy.

Wright, Charles: Richards' chief engineer on the *Hecate*. Arrived on the *Hecate* in 1861 and remained as chief engineer.

Yak-y-koss: A young girl from Quatsino whose skull had been shaped according to custom. Although it is not clearly stated, it is likely that one of the sketches taken from Mayne's *Four Years in British Columbia* is a likeness of this girl.

APPENDIX D

Glossary

Altitude and Azimuth Instrument: a surveying tool used to measure horizontal and vertical angles (or altitudes and azimuths).

Artificer: a craftsman or a skilled industrial worker.

Bower: an anchor carried at the bow of a ship.

Cable: a rough distance measurement of about 100 fathoms or 600 feet (183 metres).

Calico: a plain-woven, unprinted textile made from unbleached, and often not fully processed, cotton.

Catarrh: inflammation of the mucous membrane of the nose and air passages. Often a result of the common cold.

Chronometer: an instrument used in measuring time and, specifically, one adjusted to keep accurate time in all temperatures. It was first used in 1736 to determine longitude accurately. See also *Rates*.

Cross bearing: a line of position of several navigational aids to determine the position of the ship. Richards wrote this as *X* or *X*mc which means cross bearing magnetic course. The mc indicates that Richards was utilizing magnetic north and not true north to determine these lines of position.

Cutter: a smaller, broader boat for rowing or sailing, often used in survey expeditions. This single-mast boat is known for being seaworthy.

CWT: hundredweight; in Britain, this was equal to 112 pounds or 50.8 kilograms. In the United States, this same notation was equal to 100 pounds.

Dinghy: a small rowboat, often an extra boat on a ship.

Fathom: a measure of length containing six feet. Generally based on the length of a man's arm span and used chiefly in measuring cables, cordage, and the depth of navigable water by soundings.

F&C (Full and Change): also known as the "Establishment of the Port." This is the technical expression for the time that elapses between the moon's transit across the meridian at new or full moon at a given place, and the time of high water at that same place.

Fearnought: a thick felt used to cover the outside door of the powder magazine, portholes, and hatchways during battle. Used by Richards' crew in the temporary repair of the damage caused to the ship by running aground. Sometimes called dreadnought screen. Also a stout woollen cloth used as clothing in cold weather.

Galley: a vessel propelled by oars, with or without masts and sails. One of the small boats carried by a man-of-war ship.

Gig: a long, light rowboat, generally clinker-built, with overlapping planks or boards, and designed to be fast; or, a light, two-wheeled, one-horse carriage.

Grant's Galley: a distillery apparatus for obtaining fresh water from seawater.

Hawser: a large rope used for towing or mooring a ship.

Knot: a unit of speed, one nautical mile per hour.

Launch: a large flat-bottomed boat, larger than a scow, used as a barge to load coal at Nanaimo.

Levant: point of sunrise.

Lighter: a large flat-bottom barge, especially one used to transport goods over short distances.

Nautical Mile: a distance measuring 1.853 km.

Neaps: neap tides; tides that occur during the moon's quarter phases. At this time, the sun and moon work at right angles, resulting in smaller differences between high and low tides. Because the gravitational forces of the moon and the sun are perpendicular to one another (with respect to the Earth), neap tides are particularly weak tides.

Oakum: a preparation of tarred fibre used in shipbuilding for caulking or packing the joints of timbers in wooden vessels and the deck planking of iron and steel ships.

Observation Point (OB): a fixed point on land, for which latitude and longitude have been determined. These points are used to triangulate or determine the relative position of another point by measuring the angle between it and two observation points. Using trigonometry and the measured length between the two observation points, the other distances in the triangle can be determined, resulting in precise measurements and therefore accurate charts.

Onboard: on the ship. A second form of "onboard" is used by Richards. It means to keep the object close. In each case he states how close an object should be kept in order to avoid shallow water.

Pinnace: Richards' pinnace was a thirty-two-foot boat that was fitted by his crew in 1860. It was used to survey in advance of the ship for safety purposes.

Rates: "observation of rates" or to "rate the chronometer" means to compare a chronometer's daily loss or gain of time with true time. This was done by making observations of the sun when it was at the same altitude, once in the morning and once in the evening, or at equal altitudes. By comparing the time lapse between equal altitudes, half of which should be noon, and the chronometer's record of noon, one could determine the accuracy of the chronometer and make the required corrections. Having an accurate record of the time was vital to determining longitude.

Royal Engineers (RE): The Royal Engineers, or "Sappers," arrived on the coast in 1858. This group of approximately two hundred men played an important role in shaping British Columbia. This elite group built docks and roads and cleared the way for the cities that were to follow. By improving transportation routes, they opened the interior to prospectors and settlers. Between 1858 and 1863, they planned towns, surveyed lands, built roads and erected boundary monuments. In 1863 it was decided that the colony could no longer afford their services, and in November 1863 the people of New Westminster gathered at a farewell dinner for the R.E. officers and their leader, Colonel Moody. Many remained in British Columbia after their work as Royal Engineers ended.

Sappers: see *Royal Engineers*.

Schooner: a sailing vessel with fore-and-aft sails on two or more masts with the forward mast being no taller than the rear masts.

Shark: a smaller boat used by Richards and his crew in detached service.

Skow: a large flat-bottomed rowboat often used as a small barge (e.g. for loading coal).

Springs: spring tides; tides that occur during the full and new moon phases. Because the gravitational forces of both the sun and the moon are involved, the result is a greater difference between the high and low tide, and thus, high tides are relatively high and low tides are relatively low.

Trap Rock: any dark-coloured igneous rock that is fine-grained and columnar in structure, especially basalt. From the Swedish *trappa* for "stair," referring to the often stair-like appearance of its outcroppings.

True Bearing (TB): the direction of an object relative to true north.

Weigh: to lift the anchor, from Old English *wegan* meaning "to carry or move."

Whaleboat/Whaler: Richards used three whalers: small boats used in surveying the smaller bays and inlets during detached duty from the main ship.

Xmc: see *Cross bearing*.

References

Akrigg, G.P.V. and Helen B. Akrigg. *British Columbia Place Names.* 3rd ed. Vancouver: UBC Press, 1997.

———. *British Columbia Chronicle, 1847–1871: Gold & Colonists*. Vancouver: Discovery Press, 1977.

Akrigg, Helen B. *Dictionary of Canadian Biography Online, 1891–1900 (Volume XII)* s.v. "Richards, Sir George Henry," www.biographi.ca/index-e.html, 2000.

Barrett-Lennard, Charles Edward. *Travels in British Columbia.* London: Hurst and Blackett, 1862.

Belcher, Henry. *Narrative of a Voyage Round the World Performed by Her Majesty's Ship* Sulphur *during the Years 1836–1842: Vol. I & II.* Elibron Classics Replica Edition, Adamant Media Corporation, 2005.

Berton, Pierre. *The National Dream: The Great Railway, 1871–1881*. Toronto: McClelland and Stewart, 1970.

Bowen, Lynne. *Three Dollar Dreams.* Lantzville, B.C.: Oolichan Books, 1987.

Boyd, Robert T. *The Coming of the Spirit of Pestilence: Introduced Infectious Diseases and Population Decline among Northwest Coast Indians, 1774–1874.* Vancouver: UBC Press, 1999.

British Colonist. Various dates, 1859–1862.

Channer, Donal. Personal communication, 2008.

Cook, Andrew S. "The Publication of British Admiralty Charts for British Columbia in the Nineteenth Century," in *Charting Northern Waters: Essays for the Centenary of the Canadian Hydrographic Service*, 50–73. Montreal & Kingston: McGill-Queen's University Press, 2004.

Dawson, L.S. "The Imperial Library," in *Memoirs of Hydrography. Part II – 1830–1885*. Eastbourne: Henry W. Keay, 1885.

Deur, Douglas. "Plant Cultivation on the Northwest Coast: A Reconsideration," *Journal of Cultural Geography*. 19(2), 9–35, 2002.

Douglas, James to Henry Labouchere, Secretary of State for the Colonial Department, October 20, 1856. Transmitting census of the native Tribes of Vancouver's Island, TNA, CO 305/7.

Drucker, Philip. "The Northern and Central Nootkan Tribes," *Smithsonian Institution Bureau of American Ethnology*, Bulletin 144. Washington: United States Government Printing Office, 1951.

Duff, Wilson. Wilson Duff Collection from the British Columbia Museum, BCA GR-2809. n.d.

Fisher, Robin. *Contact and Conflict: Indian-European Relations in British Columbia, 1774–1890*. 2nd ed. Vancouver: UBC Press, 1992.

Galois, Robert. *Kwak̲wak̲a'wak̲w Settlements, 1775–1920: A Geographical Analysis and Gazetteer.* Vancouver: UBC Press, 1994.

Goodwin, Doris Kearns. *Team of Rivals.* New York: Simon and Schuster Paperbacks, 2006.

Good, John Booth. *The Utmost Bounds of the West: Pioneer Jottings of Forty Years Missionary Reminiscences of the Out West Pacific Coast A.D. 1861 – A.D. 1900.* BCA E/B/G59.

Gordon, Katherine. *Made to Measure: A History of Land Surveying in British Columbia.* Winlaw, B.C.: Sono Nis Press, 2006.

Gough, Barry M. *Fortune's a River: The Collision of Empires in Northwest America.* Madeira Park, B.C.: Harbour Publishers, 2007.

———. *Gunboat Frontier: British Maritime Authority and Northwest Coast Indians, 1846–90*. Vancouver: UBC Press, 1984.

———. *Distant Dominion: Britain and the Northwest Coast of North America, 1579–1809.* Vancouver: UBC Press, 1980.

———. *The Royal Navy and the Northwest Coast of North America, 1810–1914: A Study of British Maritime Ascendancy.* Vancouver: UBC Press, 1971.

Gowlland, J. T. E. Journal of H.M. Surveying Ship: Capt G.H. Richards, Vancouver Island. Gowlland, 2nd Master: From 1 January 1861–18 January 1864.

UBC Library (Microfilm) AW1 R77720 and Mitchell Library ML MSS 830/2 Item 5.

———. Miscellaneous pages from journals: 1860–1863. UBC Library (Microfilm) AW1 R77720 and Mitchell Library ML MSS 830/3 Item 8.

———. Journal of the H.M.S. *Plumper*: July 1859–December 1860; Journal of H.M. Survey Ship *Hecate*: 1 January 1861–18 January 1864. UBC Library (Microfilm) AW1 R77720 and Mitchell Library ML MSS 830/2 Item 6.

Great Britain Admiralty. Chart of Cape Scott 1862, D6831 Shelf Za, UKHO, Taunton.

———. Chart of Quatsino Sound 1860, D6833 Prep 41a, UKHO, Taunton.

Hayes, Derek William. *Historical Atlas of British Columbia and the Pacific Northwest: Maps of Exploration British Columbia, Washington, Oregon, Alaska, Yukon.* New and Revised Printing ed. Vancouver: Cavendish Books, 1999.

Jefferson, Chief. Deposition of Chief Jefferson, May 19, 1861, TNA ADM 1/5924 Y165.

King, J. C. H. *First Peoples, First Contacts: Native Peoples of North America.* London: British Museum Press, 1999.

Littlefield, Loraine. "Gender, Class and Community: The History of Sne-Nay-Muxw Women's Employment." Ph.D dissertation, University of British Columbia, 1995.

Mackie, Richard. *Trading Beyond the Mountains: The British Fur Trade on the Pacific, 1793–1843*. Vancouver: UBC Press, 1997.

Marshall, Yvonne May. "A Political History of the Nuu-Chah-Nulth People: A Case Study of the Mowachaht and Muchalaht Tribes." Ph.D dissertation, Simon Fraser University, 1993.

Mayne, Richard Charles. *Four Years in British Columbia and Vancouver Island.* London: John Murray, 1862.

———. Journal of Admiral Richard Charles Mayne: Feb. 17, 1857–Dec. 31, 1860, BCA EB M45A 1857–1860. Transcribed by Winnifreda Macintosh, 1964.

New York Times. "The American Question. Important Letter From Lord John Russell." *New York Times*, Jun. 18, 1861.

Nokes, J. Richard. *Almost a Hero: The Voyages of John Meares, R.N., to China, Hawaii, and the Northwest Coast*. Pullman: Washington State University Press, 1998.

Ormsby, Margaret, *British Columbia: A History.* Vancouver: Macmillan Company of Canada Limited, 1958.

Pasco, Juanita, Brian Douglas Compton, Lorraine Hunt, and U'mista Cultural Society. *The Living World: Plants and Animals of the Kwakwaka'Wakw*. Alert Bay, B.C.: U'mista Cultural Society, 1998.

Phillips, Juliet and Richard Joscelyne (eds.). *My Dearest Birdie: Letters to Australia 1874–1886 by Richard and Jessie Gowlland.* London: Jessica Kingsley Publishers, 2007.

Pojar, Jim, and Andy MacKinnon (eds.). *Plants of Coastal British Columbia, Including Washington, Oregon & Alaska.* Edmonton: Lone Pine Publishing, 1994.

Richards, G.H. to Sir George Airy, May 19, 1874. Cambridge University Library. RGO6/7.

——— to Sir George Airy, Dec. 19, 1873. Cambridge University Library, RGO6/7.

———. *The Vancouver Island Pilot, Containing Sailing Directions For the Coasts of Vancouver Island, and Part of British Columbia: Compiled from the Surveys made by Captain George Henry Richards, R.N., in H.M. Ships* Plumper *and* Hecate, *between the years 1858 and 1864.* London: Hydrographic Office Admiralty, 1864.

——— to McKenzie, Nov. 12, 1863. BCA Add MS 2431 Box 1–Box 3, File 29.

——— to C.R. Robson, Dec. 13, 1860. TNA ADM/5761 Y144.

——— to R.L. Baynes, Aug. 17, 1860. BCA GR1309 Vol. 2.

——— to R.C. Mayne, Jun. 30, 1860. BCA GR1372.

——— to R.L. Baynes, Aug. 21, 1859. BCA GR1372.

———. Hydrographic Private Letterbook, *Plumper* and *Hecate*. Apr. 2, 1857–Aug. 1, 1862. Private Collection.

Ritchie, G.S. *The Admiralty Chart: British Naval Hydrography in the Nineteenth Century.* Edinburgh: Pentland Press, 1995.

Robson, Charles R to Thomas Maitland, May 20, 1861. TNA ADM 1/5924 Y165.

Sale, T.D. *St. Paul's Anglican Church, Nanaimo, B.C.: A History, 1861–1986.* Nanaimo: Anglican Church, 1986.

Sandilands, R.W. "The Role of the Hydrographer in Coastal Nomenclature." *Onomastica Canadiana,* Dec. 1983.

Snuneymuxw First Nation Elders. Personal communication with Dr. Loraine Littlefield, 2009.

Sproat, Gilbert Malcolm. *The Nootka: Scenes and Studies of Savage Life.* Charles Lillard (ed.). Victoria, B.C.: Sono Nis Press, 1987.

Stark, Suzanne J. *Female Tars: Women Aboard Ship in the Age of Sail.* Annapolis, Md.: Naval Institute Press, 1996.

The Times (London). Obituary of Admiral Sir G.H. Richards. *The Times*, November 17, 1896.

Theodore, Mary. *Heralds of Christ the King: Missionary Record of the North Pacific, 1837–1878.* New York: P.J. Kennedy & Sons, 1939.

Thompson, Laurence C. and M. Dale Kinkade, "Languages," in *Handbook of North American Indians.* Wayne P. Suttles, ed. pp. 30–51. Washington: Smithsonian Institution, 1990.

Turner, Nancy J. Personal communication, 2010.

Turner, Nancy J. *The Earth's Blanket: Traditional Teachings for Sustainable Living.* Vancouver: Douglas & McIntyre, 2005.

Twigg, Alan. *First Invaders: The Literary Origins of British Columbia.* Vancouver: Ronsdale Press, 2004.

Vancouver, Captain George. *The Voyage of George Vancouver, 1791–1795: Vol. II.* W. Kaye Lamb (ed.). London: The Hakluyt Society, 1984.

Walbran, John T. *British Columbia Coast Names, 1592–1906: Their Origin and History.* Vancouver: J.J. Douglas. Facsimile Reprint of 1909 Edition, 1971.

Wallace, Richard William. "Charting the Northwest Coast 1857–1862: A Case Study in the Use of 'Knowledge as Power' in Britain's Imperial Ascendancy." Master's thesis, University of British Columbia, 1993.

Washington, John. Survey Instructions to Captain G.H. Richards, March 16, 1857. BCA GR-0284 Box 1, Item 14.

Yeomans, Donald K. "Great Comets in History." Jet Propulsion Laboratory/ California Institute of Technology, Electronic Document: http://ssd.jpl.nasa.gov/?great_comets, 2007.

About the Editors

Linda Dorricott and Deidre Cullon are researchers and writers with academic backgrounds in anthropology. Throughout their careers they have worked with more than twenty Vancouver Island First Nations. Their research supports treaty negotiations, aboriginal land and resource use, archaeology, the development of First Nation libraries and legal actions. The editors work as much as possible with primary sources. They believe a story is best told by the person who lived it, and that the careful reading of a handwritten manuscript, almost an anomaly in our era, may be the closest we come to hearing the past spoken. The editors were raised on Vancouver Island and live in Nanaimo. They have a special interest in the Island's colonial history and how colonial and aboriginal cultures interacted, understood and misunderstood each other. Captain Richards' journal richly illustrates such an interaction.

The journal was obtained during a research trip to England in 2006 and its publication was made possible with the authorization and assistance of the owner, Donal Channer.

Index

Québec, Canada
2012